The Talking Ape

THE TALKING APE

Keith Laidler

STEIN AND DAY/*Publishers*/New York

First published in the United States of America in 1980

STEIN AND DAY/Publishers
Scarborough House
Briarcliff Manor, N.Y. 10510

Laidler, Keith
The talking ape.

1. Orangutan—Psychology. 2. Human-animal communication. 3. Mammals—Psychology. I. Title.
QL737.P96L34 1980b 599.88'42 80-5388
ISBN 0-8128-2731-7

List of Illustrations

The Talking Ape

1

We all have vague dreams of achieving the impossible. Whether we want to be the first person to perform a solo ascent of Everest without oxygen, or to be crowned Goldfish-swallowing Champion of the World, there is usually some vision or fantasy that helps us keep our noses to their respective grindstones through the more mundane business of everyday life. Few of us manage to realize such daydreams, and fewer still find that their fantasies—once achieved—have been worth the effort.

I was beginning to think that my own personal dream was, slowly but inexorably, falling into this category. After all standing at a sink washing three dirty diapers in cold water could not be *anyone's* idea of paradise, especially with a small and very vocal infant tugging incessantly at one's leg for attention. I knew exactly how a harassed housewife must feel at the end of a busy day, but even she would not have my problems. Her infant might drag at her skirts, but at least it could not swarm up her legs with the easy grace of an Olympic athlete. Mine could.

Cody was an orangutan, just one month old and more active and mischievous than a whole battalion of human infantile delinquents. But he was no ordinary member of the species *Pongo pygmaeus*. Most orang babies spend their childhood dangling from their mother's long red fur as she moves through the high canopy of a Bornean or Sumatran rain forest; or, if they are less fortunate, they grow up isolated from all contact in a zoo cage, far from the exciting sounds and smells of a tropical jungle. But Cody was unique: he had been raised in surroundings that approximated very closely to a typical human environment, and I spent most of my waking hours observing the way in which he responded to this milieu.

At that time Cody was very interested in a blue play cube, pulling my face down to his, all the time staring hard, almost holding the object in both hands and turning it over studiously, his eyes carefully taking in each edge and face. I watched him just as carefully, noting down each movement of his action sequence. I had set myself the task of following the development of my adopted child, and the ways in which he manipulated toys and made use of them in his game was an important aspect of this behavior, one which is probably crucial to human development.

The human infant is at first a bundle of uncoordinated movements and ear-piercing screams; yet gradually each of us develops into much more elegant and independent creatures, sure of our actions and opinions. The foundations of all our behavior are, to a great extent, laid down in the first twelve to fifteen months of life, and it was this period of Cody's existence that I had decided to follow and chart. It is strange to think that all the actions we now take for granted—using a spoon to stir our tea, picking up a paper, catching a ball—had to be laboriously learned and practiced to perfection. And if Cody was anything to go by, at this early age both practice and learning are far from boring. He seemed fascinated by the play cube, transferring it from hand to hand, mouthing it occasionally (to see if it was edible?), and moving it in his grasp so that at one moment it was at arm's length and the next almost touching his eyeball. Observing Cody made me realize that we adults soon forget just how interesting a cube can be—if you have never seen it before, that is. *Everything* is new to the child, and his sense of wonder at the commonplace puts us ultrasophisticated grownups to shame.

Cody had a lot more to occupy himself with than most captive infant orangutans. He had everything that a human baby needs or requires: a crib in the brightest corner of the nursery, a baby bath, a pacifier for comfort, a baby walker, high chair, and a host of toys such as peg boards, mobiles, and rattles. All these accoutrements of the infant *Homo sapiens* were very necessary, for how can one compare the intelligence of ape and man unless one gives the growing ape as stimulating an environment as that that a human infant enjoys?

In most human families, a baby is the center of attraction. It is picked up, petted, and talked to by relations, by friends, and even by old ladies in the street. Toys are showered upon it, and any minor achievement, like holding a toy in its hand, or saying "Gloop" are acknowledged with rejoicing. If anything, the infant ape, especially if it is zoo-born and rejected by its mother, lives in a duller, impoverished environment. It is usually placed on a white cot in a featureless room with no toys or other amusements for it to look at or play with. It is hardly ever held or stroked, and it is rarely spoken to. The only stimulation it can look forward to is a rather impersonal handling three or four times a day during feeding. Small wonder that many captive apes grow up dulled and seemingly stupid. How can they be otherwise if all their lives they have never had the opportunity to explore and investigate?

I was convinced that, given the right conditions and the right education, my orangutan baby would reveal a hitherto undreamed of capacity for many skills, including, I hoped, the ability to communicate his ideas and desires vocally. In other words, I felt sure that Cody could be taught to "speak."

As if in answer to my unspoken thoughts, Cody left his toys, and crawled over to my chair. He pulled himself to a standing position and gazed earnestly into my face.

"Kuh."

He was using the sound I had taught him as a request for milk. (It rhymed with "Ugh." "You want kuh?" I asked.

"Kuh. Puh," Cody responded, and began to climb on to my knee. What did he want? From the expression on his face, he was obviously intent on something. But what?

Seating himself on my lap, Cody tried again. "Krooh," then a long pause during which his lips and tongue worked mutely. "Puh."

I was genuinely worried by this time. Whatever he had in his head, Cody was obviously very agitated by it. "What is it, Cody? What are you trying to tell me?"

He was watching my lips! Each time I spoke, his eyes locked on to my lips, and he watched with single-minded concentration until I had finished. Then it seemed he tried very hard to emulate me. It was an uncanny experience; I felt as if I were in a dream, and my scalp began to tingle with a mixture of fear and excitement. Cody tried again, putting his hands around my neck and looking desperately, into my eyes.

"Kluh. Sheesh, Pauh." With every sound, the young ape's little body seemed to tighten and twist with frustration. I was just as frustrated.

"Tell me, Cody. Speak to me."

But it was no good. Cody suddenly flopped back in my lap; his eyes were dulled, and there was a look of utter dejection on his brown wrinkled face. Without another sound he climbed to the floor and then up into the cot. Draping his blanket across his shoulders, he gave me a final, dejected look and lay down to sleep.

Cody may have been crestfallen, but I was thrilled by his behavior. It seemed that the long hours of training had at last paid off. Everything indicated that Cody had been trying to use the sounds I had taught him to signify something: to use them not simply as noises without meaning, but as verbal signs. I felt sure it was going to work. Cody was going to speak.

2

Throughout my zoological training, although it was made clear that for most purposes man could be regarded as no more than a glorified chimpanzee, it was always implicit that humans were, in fact, something special. Even the words we used were blatantly *Homo*-centric. Our lessons in comparative anatomy, endocrinology, and the rest detailed the way we had evolved from "lower forms," among which our closest ancestors were numbered. They were "lower," and we were "higher"; in fact our species, *Homo sapiens* was the highest form there was, the pinnacle of creation. In fairness to my lecturers, this attitude was probably as unconscious as it was inevitable.

Man, and western man in particular, has always regarded himself as somehow unique, a semidivine creature mystically set apart from, and with power over, the rest of the animal kingdom. This secular concept was given religious approval from very early on. It was found in the first book of the Bible where, having created Adam and Eve, God (Yahweh) sends them forth into the world with the following admonition:

> Be fruitful and multiply, and replenish the earth and subdue it;

and have dominion over the fish of the sea, and over the fowl of the air, and over every living thing that moveth upon the earth.

But are we, in fact, so special? It seems we have an inordinate desire to believe ourselves so. Even after the Renaissance the notion that our species was unique and divinely favored persisted. Man was said to have a "rational soul," which distinguished him from the brute beasts. The French philosopher René Descartes stated that animals were simply "*bêtes machines*": unfeeling, unthinking robots to be disposed of according to our whims. Under the influence of this doctrine, dogs were nailed by their paws to wooden tables and dissected alive, without anesthetic, during anatomy lessons. They screamed and howled as they were dismembered, but the practice was not thought to be cruel, for, if animals could not think or *reason*, how could they possibly feel pain?

When reasoning became defined as "the ability to learn from past experience," poor *Homo sapiens* was in something of a quandary. Even the lowly planarian worm could be taught to link the shining of a light with the sensations of an electric shock. If a light was shone on the unfortunate worm before the shock was administered, the planarian would eventually learn from its past experience and react to the light as if it were an electric shock. Did this mean that we were no better than a simple invertebrate—more complex perhaps—but essentially the same? The prospect was almost too ghastly to contemplate, but the human race was saved from this horrible, ego-shattering humiliation by science. It was decided that all we needed to do was to look at man, see what he was capable of, and then see if animals could do the same. If they could not, *that* became the new criterion for distinguishing the Lord of Creation from his subjects. This "scientific method" has led to the rather laughable spectacle of man proudly hoisting the flag of uniqueness on innumerable occasions, only to furl it again quickly as some new discovery in animal behavior renders it inappropriate.

One of the first of these scientific definitions of man was "man the tool user." According to this theory, man was the only creature now living that could utilize inanimate objects as tools. Unfortunately, no sooner was the criterion established than cases of tool use in animals other than man began to appear. There was the Galapagos finch, *Cactospiza pallida*, a bird that regularly uses cactus spines held in its beak to dig out insects from the crevices in tree bark. In the southern United States another bird, the brown-headed nuthatch, *Sitta pusilla*, uses a similar tool, a fragment of bark, for the same purpose. Birds as widely distributed as the black breasted buzzard, *Hemirostra melanesterna*, of Australia, and the Egyptian vulture, *Neophron percnopterus*, use rocks to break open eggs, the former by dropping a stone into the victim's nest, the latter by hurling a stone at the egg. The list of tool users among the primates is almost endless. Benjamin Beck, writing in 1975, mentions 16 species that are known to use tools, including all the great apes and both New and Old World monkey species.

It was very unwise to choose tool use as distinguishing man from the rest of creation. In fact, the theory should never have been proposed. Had the proponents of man's uniqueness studied the subject diligently, they would have found numerous references to this topic in much of the early literature of natural history. In *The Malay Archipelago*, published in 1869, the great British naturalist, Alfred Russel Wallace, mentions that the black cockatoo, *Probosciger aterrimus*, cracks open nuts with its beak while holding the nut sandwiched inside a leaf. The cockatoo apparently uses the leaf to get a better grip on its food in much the same way as we hold a bottle top with a towel for ease of opening. This observation is probably the first published record of tool use in animals other than primates.

So, the human race was not alone in using tools; another definition of man was needed. It was not long in coming, and one could almost hear the sighs of relief. Man was the only creature that *made* tools. This seemed a far better bet, but once again, animal studies spoiled the neatly written story. The pioneering ethologist,

Jane Goodall, observed chimpanzee behavior that was strikingly similar to human tool making. The chimps use twig tools to obtain protein-rich termites during certain seasons of the year. The way they do this has been called, very aptly, "fishing." What the fishing chimpanzee does is to select a branch or twig from a bush, snap it off, and then remove the leaves so that what is left is a long, thin wand. This wand is then pushed into the termite nest. Responding to the assault on their clay citadel, the termites attack the twig, seizing it with their jaws, whereupon the chimpanzee withdraws his tool, complete with a row of termites, and eats them.

The important thing about this behavior is that the chimpanzee actually *modifies* the object that acts as a tool, both by breaking it from the bush and by stripping the branch of its leaves so that it can be slid more effectively into the termite mound. Chimps have also been seen to modify leaves by chewing them to make a spongelike mass in which to collect water, and to make the end of a branch thinner so that it will fit between the lid and base of a box of bananas. Even the despised orangutan has been seen making a tool of sorts. A male orang, held in a large battered cage, plaited straw to make a rope, which he then hung over his cage bars and used as a swing. All this evidence showed clearly that tool manufacture was just as obsolete as tool use as far as uniqueness was concerned. Those who wished to differentiate man from the beasts were forced to cast around for another definition.

At present, it is "man the speaker." Man, we are told, is the only creature to possess a language. But just what is a language and how exactly does it differ from the communication of other animals? Other creatures' systems of communication are emotional: their cries are of fear, rage, and hunger, and their calls are made whether there are conspecifics in the vicinity or not. Man, by contrast, usually directs language toward other humans; he is able to converse with his peers about things in his environment, to tell them not simply that he is afraid, but that he is afraid *because* he has seen a huge lion which is behind the nearest bush to the right and that the lion is almost ready to attack. Human language is thus a far more precise method of communication than a simple call of fear. Further, it allows man to scheme with his fellows and to plan in detail how they can defend themselves from the lion.

Essentially, then, language is nonemotional; it allows our species to step outside itself, to name objects and events in the environment, and to speak of them as having existence in the past and the future, as well as the here-and-now. For humans, language is a communication system *par excellence.* It is extremely rapid. If you compared the time it would take to tap out this page in Morse code with the time it takes to read it, you would find that the "reader" had finished the page before the "tapper" had even reached the fourth line. The coded sound series of language conveys information at least ten times faster than any other method of signaling; and for a hunting animal like *Homo sapiens* this is a great advantage. Human language is by far the most efficient of any system of communication and, as far as we know, no animal possesses a human-type language in the wild. But some scientists have gone beyond this, and claimed that language is species specific—that is, that language is the sole preserve of the human race. They believe that man's brain is in some way programmed for this system, and, *ergo,* no other creature can learn, or even be taught, a language. The implication is obvious: man's position as head of the table is once again established. As the naturalist Gaylord Simpson has declared: "Language is the most diagnostic single trait of man; all normal men have language; no other living organisms do."

Even as an undergraduate I found this, and similar statements, not quite believable. To be sure, no other species talks to its fellows about the weather, or discusses England's chances in the latest test match, but I was not convinced that such a huge abyss existed between human language and the so-called systems of other animals. Certain creatures are known to say more than "I am afraid," etc., and some come fairly close to bridging the language gap.

In the forest margins of East and Southern Africa lives a species of monkey called *Cercopithecus aethiops*, the vervet monkey. Until 1967 nothing particularly interesting was known about this primate; then a study by Thomas Struhsaker of the University of Chicago revealed that it had not one, but three different alarm calls. The first, a "r-raup" sound, is given on the approach of an aerial predator, such as a monkey-eating eagle. The second sound is a chuttering call used to warn of snakes, while a "shirp"—the third sound—designates ground-living predators and will send the vervets scattering to the very tips of the branches for protection. What is interesting is that all members of the troop respond appropriately to the different calls—they understand that each call is specific to a particular referent, be it eagle, snake, or cat. If the vervets had learned these three calls (they are thought to be unlearned, instinctive responses), then we would have a *protolanguage*, perhaps similar to that used by our ancestors millions of years ago.

But there is no need to travel as far as Africa to find animals that show evidence of protolanguage. After reading about Struhsaker's vervets, I realized that a creature much nearer home was showing communicatory behavior far in excess of simple emotional screams and cries. That animal was the domestic dog. During my teens I was inseparable from a mottled brown, shaggy-haired, crossbreed dog named—rather unoriginally—Laddie. I say we were inseparable, but in fact Laddie was a tramp. He was a true and trusted friend for as long as there were no receptive lady dogs in the neighborhood. When there were, Laddie would disappear for several days, sometimes weeks, only returning when he had been either victorious. or pre-empted by some equally determined suitor. The loyal side of his nature would then hold sway—until his nose led him to another lady love.

When he was at home, Laddie made up for his periodic disappearances by affording us a view into the highest levels of the canine mind. He really was a most intelligent dog, and some of his behavior was positively human. I know, of course, that this is a favorite theme of devoted pet lovers, but perhaps there is some truth in all their eulogizing. How else can one explain Laddie's behavior in the case of the missing Thermos?

We had been on a day-long pigeon shoot during the course of which we had traversed several fields and numerous woods. Naturally, we had taken lunch along with us, including a nice flask of coffee. As we were about to set off for home we noticed that the Thermos flask was missing. We immediately repaired to our picnic site, but there was no sign of the flask. As we scoured the surrounding grass and brambles, Laddie also disappeared. Ten minutes later, as we were about to conclude that we had lost both flask and dog, there was an excited bark, and the missing hound stuck his shaggy head through the nearest hedge, his eyes glinting with excitement. He barked, then disappeared. A short time later the procedure was repeated. At Laddie's third display, I realized he was asking me to follow.

He led me toward a wood, rushing ahead sometimes, only to backtrack and give me a bark of encouragement before heading off once again into the underbrush. Eventually, after about half a mile, I stepped out into a small glade to find Laddie standing proudly over the lost flask. Clearly the dog had virtually asked me to follow him to where he had made his find. The strange thing was, and is, that we had only talked among ourselves about the flask, and not one posed Laddie the problem of finding it. He had worked out for himself what was wrong. So, even if you cannot accept that Laddie "asked" me to follow him, you must surely grant him a limited understanding of the English language.

You hear similar reports every day. Dogs pulling on people's sleeves to lead them to an injured master, dogs scratching at the door then looking round pointedly to convey the message that they wish to be let out—everyone knows of similar cases. Even more intriguing, though ignored because of its ubiquity, is a dog bringing its leash to its owner in a mute request for "walkies." While it is not human language, is the dog not using his lead in a sort of "sign language" where the leash comes to symbolize the act of walking? I began to wonder just what might be achieved if we could give dogs sounds for different activities.

Other people were obviously thinking along the same lines and even going a stage further. During my final year at the university I read of an experiment that set my thoughts racing. It was being conducted by two American psychologists, Robert and Beatrice Gardner of the University of Nevada at Reno. Their thoughts were focused, not on the domestic dog, but on man's nearest living relative, the chimpanzee. The Gardners believed that this animal could be taught a gestural language. They based their belief on the findings of such researchers as Jane Goodall and Adrian Kortlaandt, who had reported that the chimp's behavioral repertoire contained many gestures, some of which were so close to human signs as to be indistinguishable. Chimpanzee acquaintances who have not seen each other for some days react in a strikingly "human" fashion. They approach each other rapturously, embracing and even kissing, showing every indication of affection. Again, a chimp that spies an acquaintance with some desired morsel of food will often follow the food owner, holding out its hand in a typically human begging gesture. Chimpanzees have been seen to pat the heads of their infants and even tickle them under the chin in a manner that is strongly reminiscent of our own. The chimpanzee as a species was obviously capable of generating, recognizing, and responding to a wide range of gestures. This ability led the Gardners to believe that the species would be an ideal subject to learn a human sign language.

Human sign language has existed since before the beginning of recorded history. The Plains Indians of the United States, native tribes in Africa and India, and the oboriginal peoples of Australasia all have, or had, a system of gestures by which they could communicate silently with other members of their tribes, especially during the final stages of a hunt. For example, a Bushman in the Kalahari desert of South Africa will communicate the presence of a nearby hare using a sign which is very similar to the Churchillian "victory" symbol. This is a great advantage in situations where speech would alert the intended victim, and we can speculate that early man may have had developed sign languages before he learned to speak. But the advantages go far beyond the hunting situation. A deaf child in a more "civilized" society is at a great disadvantage; almost all our discourse is carried out using spoken language. This is not the case in signing communities: with sign language a deaf aboriginal child can grow and assimilate most of his tribe's traditions and lore and also enter effectively into many day-to-day events.

Sadly, most of these traditional sign systems are dying out as civilization moves in upon the scattered remnants of primitive peoples. However, another sort of sign language is far from extinct—the system taught to deaf people as an alternative to spoken language. There are several variations on this gestural language throughout the world, the "species" found in the United States being American Sign Language or Ameslan (ASL).

ASL is very widespread among the deaf of North America. It consists of about 1600 different signs, some of which are completely arbitrary in their meaning, while others have an obvious derivation (for example the sign "to drink" is made by bringing the fisted hand, with thumb extended, to the mouth as if drinking from a cup). ASL allows those deaf people who understand it to lead a relatively normal life, and it is regarded as a true language by many authorities. There are some purists who argue that language must necessarily be *spoken* language; but this is a rather narrow view, and certainly a large number of deaf signers would be very annoyed to be told that they did not possess a language. It was the ASL system the Gardners chose for their pioneering study.

Their experimental subject was a wild-caught, female chimpanzee, aged between ten and fourteen months, and named Washoe after the locality in which the study was carried out—Washoe County, Nevada. The Gardners' technique was to immerse Washoe totally in an ASL environment. Throughout her infancy she was surrounded by ASL signers—no spoken English was allowed within earshot of the young chimpanzee. While, for the human infant, virtually all intercourse with the world of men is via the spoken word, for Washoe the medium of information exchange was to be ASL. The only sounds allowed were spontaneous ejaculations of pleasure, surprise, or annoyance. As well as the total immersion environment, through which the Gardners hoped Washoe would "naturally" pick up ASL gestures, the young chimpanzee was also deliberately taught signs. Several were taught by rewarding Washoe with something she liked whenever she made a movement similar to a sign, but by far the most efficient method was by leading the infant's hands through the signing movement, a technique called "guidance." Although Washoe often received food rewards for her signing, tickling—which all chimpanzees crave—was found to be far more effective.

With the guidance-tickling method the training program succeeded beyond the Gardners' wildest imaginings. Washoe took quickly to the new language, learning 21 signs in the first four months, and using them, in most cases, correctly. But what was even more interesting, she *combined* them to form mini-sentences. This began as soon as she had a "vocabulary" of eight signs. Phrases like "food please" and "tickle me" began to appear regularly. Washoe also signed spontaneously when alone, as if talking to herself. Intrigued by these findings, I wrote to the Gardners, and within a few weeks they sent me further details, including month by month summaries of Washoe's progress. The chimpanzee was making quite fantastic advances with ASL. Her vocabulary of signs was increasing steadily, and with it the length of her sign combinations. On occasions when a very eager Washoe was impatient to be taken outside for a walk, combinations such as "you me go out there hurry" were signed to her companions. Washoe was also beginning to generalize signs: after learning the sign for hat, during week 103 of the experiment, she put a plastic bag on her head and then rushed around the mobile home in which she lived signing "hat, hat." A few signs were even created by the chimp—like bib, which Washoe wanted but had no sign for. She sat stymied for a while, then drew the outline of a bib on her chest. This was a sure sign that she knew her gestures represented specific referents. The Gardners later found that this chimp-created sign was, in fact, identical with its ASL counterpart: Washoe's interchanges with her companions had also begun to take on a humanlike quality. The following conversation was recorded during week 111 of the experiment when Washoe was offered a cherry to eat:

Washoe: Gimme.
Companion: What do you want (signed by "What this?")?
Washoe: Food. Gimme.
Companion: Ask politely.
Washoe: Please.

The exchange is remarkably similar to a mother talking to her child. And, like a human child, Washoe was adept at using her signs for devious ends, as when she signed "open" at a cupboard. When her request was complied with, she grabbed a handful of forbidden objects from the cupboard and beat a hasty retreat. A few minutes later she returned and signed "open" again. Her companion refused. Washoe then went to her potty and, after much effort, produced the tiniest amount of urine. Her companion had then to open the closet to dispose of her urine, giving Washoe the chance to grab yet another handful of forbidden objects.

Even more unusual was the striking resemblance that Washoe's signing showed to the development of spoken language in human children. Roger Brown had shown in 1970 that the words of children's early sound combinations could be divided into two categories, "pivot" and "open" words. The pivot words seemed to act as verbs, the open words as nouns. As a result, although pivot words were very frequent in occurrence, there were always far more open than pivot words. A pivot word might be "go," and Brown found that it would be used in combination with several "open" words, in phrases such as "go sweetie" (when candy was desired), "go shops" (to be taken to the shops) "go potty" (when the child wished to relieve itself). An almost identical classification could be made for Washoe in her early stages of ASL acquisition. One of Washoe's pivot signs was "gimme" and she used it with numerous other signs, e.g. gimme sweet, gimme tickle, etc.

At that time, just four months before my finals, this was as far as the Gardners' fascinating experiment had progressed. Further research, over the past five years, has greatly extended the study and shown that the chimpanzee has far more linguistic potential than anyone believed. The first question—would other chimps prove capable of learning ASL?—has now been answered. Washoe is *not* some kind of chimpanzee Einstein. All chimpanzees studied so far have shown a great aptitude for ASL, and some, being raised from birth with experienced ASL "speakers" have displayed a much greater rate of uptake than did Washoe. Two young chimps, Booee and Bruno, have been observed to sign between themselves; they usually do this when in need of mutual comfort, "come hug" being a favorite request. On one occasion Booee approached and signed "tickle Booee" to his friend. But Bruno, far more interested in finishing off the apple he was consuming, refused this request with "Booee me food."

With their novel conquest of sign, Washoe and the new "talking" chimps have revealed that the chimpanzee apparently thinks in much the same terms as does mankind. Lacking a sign for ducks, Washoe spontaneously called them "water birds." As good a description as I can think of. When Washoe first took a bite of a radish, she screamed at the taste and immediately labeled it "hurt food."

Most interesting of all these new discoveries is Washoe's treatment of the word sign "dirty." She was taught this sign to refer to dirt or feces and to soiled items, in other words she used it as a noun. However, Washoe one day spontaneously changed the sign from a noun to an adjective, and not a very complimentary one at that. She was in the monkey room of a laboratory with Dr. Roger Fouts, previously one of the Gardners' research students. Dr.

Fouts planned to teach Washoe the sign for "monkey" by showing the chimpanzee various monkeys and then taking her hands through the appropriate movements to produce the ASL sign for this word. Before they had a chance to start, Washoe had a violent argument with a rather cantankerous macaque monkey.

> They threatened each other in typical chimpanzee manner and macaque manner . . . I stopped the aggessive interaction and turned her around so that she was facing two siamangs . . . After I had molded her hands in to the correct position three times she began to refer to the siamangs with the monkey sign . . . Next, I turned her toward the adjacent cage holding some squirrel monkeys, and she transferred the monkey sign immediately to them. After she called the squirrel monkeys "monkey" several times, I turned her around and asked her what her previous adversary, the macaque, was, and she consistently referred to him as "dirty monkey" . . . in essence it could be said that she had generated an "insult."

All this was in the future. At that time, just four months before my finals, I knew only that a chimpanzee in the United States had learned more than a score of word signs. But this information alone was terribly exciting. The thought of talking to another species was incredible. Where would it lead? Would we ever be able to question apes, to ask them how they viewed the world? What life meant to them? The idea was fantastic—almost like talking to some extraterrestrial! I knew then what I would do after the examinations. If I worked hard and got a good degree, I could virtually pick my research topic. And it would be ape language.

3

Getting a good degree is far easier than getting a chimpanzee. My first problem was finding an institute of learning that would allow me to carry out research into ape language. This was no easy task when the current fashion zoology was the so-called New Biology—the study of life at the subcellular level—and pursuing a doctoral thesis usually meant unraveling the fine structure of mouse mitochondria or some equally esoteric branch of cell biology. To many zoologists the idea of a talking ape was simply too preposterous to be taken seriously.

After many raised eyebrows and an even greater number of rejections around the country, I began to despair of zoology departments; I concentrated instead on the departments of psychology and even anthropology. It was by sheer chance that I applied to Durham University. Here, unknown to me, the head of the anthropology department, Professor Eric Sunderland, was very interested in all aspects of primatology. An interview was arranged at which I outlined my plan of study: the home-raising of an infant chimpanzee in an approximately human environment so as to allow optimum conditions for the teaching of sign language. Luckily Professor Sunderland, a tall, rather ascetic-looking academic, had heard of the Gardners' Washoe experiment, and so, instead of the usual "Is the boy mad?" response, I was met with great interest and enthusiasm for the project.

Professor Sunderland promised me all the support I might need, as long as I could overcome one very large problem. Like most other government institutions, the university was perennially short of money. Assuming I obtained the baby ape I needed, where would I keep it? The university, he said, would find the cash to purchase a chimpanzee provided I could supply accommodation, at my own expense. This really stumped me. On a student grant it was impossible to pay for the construction of the specially-built kindergarten I envisaged for the experiment. I could not ask my parents to lend me a room because at the time the house was fully occupied. I thought of rented accommodation, but numerous inquiries at student lodgings rapidly convinced me that no self-respecting landlady would contemplate a young man and an infant chimpanzee as tenants, no matter how laudable and scientifically valuable their activties might be.

The prospects were so bleak, I almost gave up hope of ever pursuing my increasingly doomed project. I even began noticing positions in electron microscopy with more than a passing interest. Then it struck me: people and organizations unused to animals such as apes might balk at the idea of keeping them, but what about establishments that catered specially for such beasts, the zoos and safari parks? Surely they would be interested? In return for accommodation, I could arrange to allow the public in to watch signing tuition at set times of the day. The zoo would have an extra attraction, a cuddlesome baby ape (and a tame scientist), free of charge. Likewise gratis would be the publicity that this sort of experiment was sure to generate—and what commercial enterprise would look askance at free column inches or TV air time? Fired with new enthusiasm, I began poring over the yellow pages in search of possible "lodgings."

As it happened, my choice of accommodation was extremely narrow; there was only one zoo, Lambton Lion Park, in the vicinity of Durham University. Fortunately, when I summoned up sufficient courage to telephone, the director of the park showed a degree of quiet interest in my proposal. An interview followed, and this time my idea was accepted—but only in principle. As with the university, there was one important proviso. It appeared that the only way to obtain a young chimpanzee in a reasonable time was to have it wild-caught and shipped to the United Kingdom. That meant quarantine, and there were no quarantine facilities at Lambton. However, the director of the park assured me that the zoo would build a quarantine center provided they were quite certain that the young chimp would be purchased by Durham University.

So there it was, a classical "double-bind." No ape until accommodation secured, and no accommodation until ape secured. Neither the zoo nor the university would move first, and unless one did, my project was doomed. The only way to break the stalemate, it seemed to me, was for each side to believe that the other had already begun work on its particular side of the bargain. If not, the position would remain deadlocked until my grant ran out! I had to do something to get the project under way. So—with fingers firmly crossed—I went to each party in turn and assured both that their opposite number was already working at top speed, the zoo to construct the quarantine station, the university to obtain the chimp.

I have no doubt that had either side telephoned the other to check that all was really as I said, I should now be selling ice cream or emptying garbage cans for a living. As it was, no one asked any embarrassing questions, and within a week I was safe. By then, both groups were working away at their respective tasks, each believing that the other had set the ball rolling. The zoo had even allocated me an assistant: a tall, blond enthusiastic keeper called Philip. Philip was to help me feed and care for the infant chimp, and I had agreed to instruct him in sign language so that, when my research came to an end, he would continue teaching the pongid. Philip was very keen on this aspect of the experiment, and I believe he looked forward as eagerly as I did to the chimp's arrival.

The quarantine building was completed three months later, just two weeks before the baby chimp was expected. It was a formidable-looking home for such a tiny occupant, more like part of some futuristic hospital than an enclosure for an infant ape. The basic module was a normal zoo cage, about seven feet wide by fifteen feet in length, fitted with the usual bars and door. Around this a second cage had been constructed, double-glazed with armored glass, and a second air-tight door. This door connected with the first by means of a short passage; the passage acted as an air lock between the controlled environment of the cage proper and the outside world, so that no nasties—viruses or bacteria—could escape to the surroundings. To enter, one had to follow a strict routine of unlocking and opening the outer door, entering the passage, closing and locking the door behind one. The cage had its own air supply; two fans introduced and extracted all air, the extracted air being sterilized before it was discharged to the outside.

The clinical atmosphere was emphasized by the clothing of all personnel entering the cage. Long rubber gowns, similar to those used by surgeons, had to be worn on all occasions. Completing the surgical outfit were white masks, green wellingtons, rubber gloves, and green pork-pie hats. All these frills would no doubt instill confidence in the efficiency of any hospice, but I had real reservations about their effect on my experiment. Looking at my vulcanized face in the glass of the cage, I began seriously to doubt the feasibility of raising the ape in an "approximately human environment." How could the ape be reared normally when faced with such a sterile "home," and so weird a selection of green-gowned "parents"?

Our baby chimp arrived right on time two weeks later. The gatekeeper telephoned to say that a "chimp van" had just passed through his checkpoint and the reception party of directors, keepers, and me rushed out into the driveway to welcome the newcomer. In contrast to the multitude of directions and regulations that had been enforced in constructing the quarantine cage, and the army of civil servants and inspectors who had marched all over the building in the hope of finding something amiss, the chimp's entrance was remarkably unceremonious.

I, for one, had had visions of a six-wheeled, specially constructed vehicle bearing slowly down on us on the quarantine station, perhaps with a red warning light rotating atop its armor-plated hull. But no. Up the drive tore a quite ordinary-looking Ford Escort van. The driver screeched to a halt, stuck his head out of the window, and demanded in a broad cockney accent, "You the lot waiting for a chimp, mate?"

I nodded my assent and watched in shocked silence as the shirt-sleeved driver pulled from the back of the van a rather tatty, tea-chest sized wooden box and carried it nonchalantly into our special-built, ultramodern, supersterile quarantine cage. No air locks, no double glazing, no nothing. He didn't even wear gloves!

With his load delivered, the cheery Londoner bade us an unconcerned "Ta-ta," leaped into his van, and sped off. In the face of such a devil-may-care attitude, both Philip and I felt rather foolish as we togged ourselves out in our protective clothing before entering the cage.

All the feelings of embarrassment vanished as soon as we opened the box. Peering into its dark interior I spied two large and very frightened eyes staring back. I bent down with hands outstretched, and, with only a little hesitation, a miniature female chimpanzee climbed into my arms.

She was absolutely beautiful. From head to toe, the baby ape was covered in long, silky, blue-black hair. Only her face, her huge gnome-like ears, and her soles and palms were clear of the long straight strands. On her face and palms the skin was soft and spongy, just like that of a young human infant to the touch. And touching, or any other sort of physical contact, seemed to be what "Lucy"—as I immediately decided to call her—wanted most of all. Chimpanzees are very sociable creatures, and, as many field studies have shown, they express themselves with hugs, kisses, stroking, and a host of other tactile movements. It has been said that a single chimpanzee is not a real chimpanzee at all—the species has an inborn need for company of its own kind. If a chimp cannot find comfort with its fellows, it will make do with whatever animal is on hand, anything from a marmoset to a man.

Lucy had been completely deprived of contact with any species since her capture in the Cameroon several weeks before, and even more than food it was membership in some sort of family or community that she craved. To this lovely little ape, the green-robed figures were better than nothing. Within a few minutes Lucy had snuggled up gently against my chest, her head down and eyes closed. She gave a great yawn, squeezed my arm as if to make sure I was still there, and then, with a contented sigh, immediately fell asleep. I felt suddenly very sorry for this poor little prisoner that my curiosity had torn from the forests of Africa. If I had known then the horrors she had endured in getting from the Cameroon to England I should have felt a great deal worse.

To be honest, I had never given much thought to how an infant chimp such as Lucy was captured. I wanted a baby ape to work on, and, at that time, I was so obsessed with the idea that the methods of capture never even crossed my mind. It was not until several years later that I discovered the horrendous facts. The capture of chimpanzee infants is accomplished only with a horrible amount of bloodshed. Native hunters scour the forest until they locate a band of chimpanzees, usually when they are resting in the heat of the day, and surround the group with nets. When the band is finally encircled, the hunters move in. They ignore most of the animals in the group and seek out the mothers with infants. The chimpanzee female is a dedicated mother, and she will allow the capture of her child only over her own dead body. This is exactly what happens.

The female is shot with a rifle or shotgun or sometimes hacked to death with a machete, and the infant plucked screaming from her corpse. On occasions the "valuable" infant is shot accidentally during a capture attempt; others, separated from the family group and held captive in strange surroundings, simply pine away and die. Still more perish because of the primitive and unhygienic transport methods that are still used to bring them to the export ports, ready for shipment to zoos and research establishments around the world. Dr. Adrian Kortlaandt, a world authority on the chimpanzee, has estimated that from four to six adult females are killed for each chimpanzee infant successfully exported. If we add to that number a figure of between one and three infants for the chimps that die in transit, then the capture of Lucy for my experiment must have cost the life of at least five, and possibly nine members of her species. Knowing the grisly facts, I will never again import any ape.

But at the time I was completely ignorant of the havoc that my request for a baby chimpanzee had wrought. All I knew was that Lucy had arrived, that she needed food and comfort, and that I was determined she should get both commodities in abundance. There was just one thing that worried me about Lucy: the young chimp appeared to be rather lethargic, and her eyes, to me, seemed just a shade too yellow. For an instant a little cloud of disquiet flitted across my mind—just what sort of disease might Lucy have brought in from the Dark Continent? It was a disquieting thought, but the feeling of unease soon passed; Lucy perked up considerably when Philip brought in her meal. From the first spoonful Lucy tucked into the plate of baby cereal with such gusto it seemed she had eaten it all her life. As I spooned the gruel into her ever-open mouth, I put all gloomy thoughts aside: after a few weeks' recuperation, Lucy would be ready to learn the first signs of her new language. With luck, three months from now, we could be "conversing" with a chimpanzee. Everything was going to be all right.

A figure dressed in green rubber gown, cap and gloves, entered the room, closed the door, and walked toward me. I remember thinking that it must be Philip, but what was I doing lying down—and where was Lucy? Then all the ghastly facts crowded back into my sleep-befuddled brain. This was not the zoo, and that was not Philip standing at my bedside. I was in the hospital, and the green garbed figure was a nurse, gowned against my infection, and come to feed me breakfast and my first two pills of the day. A lot had happened during the three months since Lucy had arrived at the wildlife park.

Lucy had not brightened up after a few good meals, as I had hoped she would. I became more and more concerned about her general state of health, especially as I could feel myself increasingly depressed and fatigued as the weeks wore on. Two months after Lucy's arrival I awoke in the early morning with terrible pains in my back and stomach and a very high fever. The doctor diagnosed an ulcer, but two days later my urine turned a very dark brown color, and I knew then that, at very least, I was suffering from hepatitis. My doctor confirmed the diagnosis, and the last thing I remember was lying on a stretcher and being loaded into an ambulance bound for Sunderland's Havelock Hospital for Infectious Diseases.

The next two weeks are completely lost—I was either delirious or unconscious all the time. When I did finally surface it was to find that my private little world of ape-rearing had fallen apart. Lucy had been sold to a London research laboratory, and as far as the wildlife part was concerned all bets were off. The park had also lost one of its keepers: Philip had also come down with hepatitis. Like me, he would be hospitalized for at least another four weeks, after which he would need a convalescent period of three months, perhaps longer. It was all very depressing, but I made up my mind that I was going to continue in this line of research. If all I was strong enough for was lying in bed and reading, then I would make sure that, when I was fit enough to continue, I would be the best-read Ph.D. student in the country.

After a tasteless breakfast of hospital poached fish, I turned to the first of three large piles of scientific papers that the university had kindly photostated and sent over. The first pile was concerned with primate vocalizations. Although this was not directly concerned with my own research topic of sign language, I thought I had better be acquainted with all types of primate communication systems.

I started on the top reference, a short paper by Mr. A. L. Garner, published in 1896. Garner was a big-game hunter who also kept a chimpanzee as a pet. One day, it seemed, the thought struck Mr. Garner that this chimp might be able to learn speech, if only someone bothered to teach it some words. For reasons best known to himself, Garner decided to teach a polyglot assemblage to the ape—the English word "mama," the French "feu," the German "wie," and a native word "nkgive." Goodness knows this latter sound was hard enough for most humans to produce, let alone a chimp; nevertheless, Mr. Garner persisted. He claimed that the chimp had attempted to imitate the words, but had unfortunately died some three months after tuition began. The thought crossed my mind that the strain of attempting to be the world's first quadrilingual ape might well have been the cause of the unfortunate animal's untimely demise. But such thoughts were not very scientific and I left Mr. Garner and started on reference number two. By a strange coincidence this paper also concerned an attempt to teach apes to talk, but it was entirely different from the previous one.

I was immediately fascinated by the account, published in 1916 by one D. Furness, a medical doctor who, at that time, was living in the depths of the Bornean jungle. Dr. Furness had possessed a pet female orangutan to whom he had attempted to teach "language." He gave the animal daily "lessons," modeling the required word and then manipulating his pet's tongue and lips and blocking off its nostrils with his hands. After six months the ape was able to voice the word "papa." Furness then taught it a "kah" sound by blocking the orang's nose and pushing a bone spatula against its tongue. Later, the ape learned to use her own finger to block her nose. After she could say "pah" and "kah," a "thuh" sound was gradually added to the orang's repertoire. Unfortunately, the poor creature died soon after this, of unspecified causes, just four or five months after learning its first sound.

I could hardly believe these results as, at the time, accepted opinion held that it was impossible to teach any ape to speak within a reasonable time. This view was based almost solely on one study, the so-called Hayes experiment of the late 1940s. Two

American psychologists, Catherine and Keith Hayes, were among the first to attempt a detailed scientific study of the vocal linguistic potential of great apes. They obtained their subject, a female chimpanzee they named Viki, at the tender age of three days, and raised it in much the same way as a human child is reared. Viki had her own room, cot, toys, and all the other accoutrements of a human child. But unlike infant *Homo sapiens,* Viki also received concentrated speech training from the age of five months until the experiment was terminated six years later.

Speech training consisted of withholding food until Viki made a sound. Then she was rewarded with a spoonful of food or a drink of milk, in the hope that she would realize that sounds could "buy" her meal. Progress to speech was extremely slow, and it was a further five months before Viki achieved what seemed for her an extremely difficult feat—the production of a harsh, staccato grunt each time her food was offered. Following this success, a "shaping" method was initiated, with Viki's mouth manipulated during sound production and only successively closer approximations to the required sound being rewarded with food. From this came the word "mama." Twenty-five months after this Viki had learned a "pa" and "kah" sound. She subsequently learned "up" plus a clicking and a "tsk" noise in her vocabulary.

To judge from cine recordings of Viki performing her "words," the sounds were not well formed, and in at least two of them Viki had to place her hand across her nose to prevent the escape of air. Human beings do this internally, raising their soft palate or velum to close off the nasal chamber, but from the evidence of the Hayes experiment the chimpanzee was not able to do this.

Such disappointing results had led to the popular belief that all apes are incapable of learning a vocal language. Yet here was Dr. Furness, over thirty years before the Hayes experiment, claiming that an orangutan had learned as many sounds as the chimpanzee Viki, *and* in only one-quarter of the time! Could it be true? Might the orangutan have an inborn predisposition to talk that the chimpanzee lacked?

The more I thought about it, the more I became determined to find out. After all, to state that it was impossible for *any* ape—

chimp, orang, or gorilla—to learn to talk when only the chimpanzee had been looked at in any detail was, to say the least, unscientific. Until the orang had been studied scientifically, no one could say what its limits were with regard to language. And more than that: Furness's teaching techniques had been rather old-fashioned; it could well be that an orang, given the same sort of tuition that Viki had been exposed to, would far outshine the chimpanzee in the production of voluntary sounds. Could I possibly repeat the experiment and improve on the result?

The research would be extremely exciting, and it was very necessary too. Recent studies on sign language in chimps, like that of the Gardners with Washoe, *were* important, but I realized that they could never tell us the way our own *vocal* language evolved. Even if our remote ancestors did use gestures to communicate, they would have had, at some time in their evolution, to change to vocal language to avail themselves of all the advantages of such a system. By teaching an infant orangutan to talk, I could gain some insight into the evolution of language in our own species, as well as charting the linguistic capacity of one of our nearest relatives.

There were other reasons why the research appealed to me. The orangutan has long been regarded as both dull and stupid: in consequence it has never been studied extensively, psychologists preferring to spend their time where it seemed most productive, on that whiz kid of the great apes, the chimpanzee. Almost everything about the orangutan was unknown, and that gave me a great deal of scope. I wanted to look at the species' growth patterns, its psychological development, how and when it learned to crawl, climb, grasp objects, and do all the things an adult orang can do. Many of these aspects of the orang's behavior could be compared with human infants to see in what way the orang differed from, or mirrored, human development. Such knowledge could greatly further our understanding of the way in which the infant, both ape and human, progressed to adulthood. It might also help work on subnormal or retarded human infants and perhaps give leads to their treatment. My mind was racing with ideas, but there was one big problem: first find your ape.

4

Apes are like people—first impressions count. Although I had been acquainted with the two adult orangs in the cage in front of me for less than ten minutes, I had already summed up their characters. As it turned out, my assessment was not very far from the mark. Six months had passed since the ill-fated Lucy experiment and, with my convalescence complete, it was time to try again. I was now some one hundred miles south of Durham, at Flamingo Park Zoo in the North Riding of Yorkshire. The two orangs were the parents of the baby ape I had finally found to study, and, looking at them, I hoped their offspring would share none of the more obvious qualities of either.

The father, Adam by name, was the less inelegant of the pair. In fact, in a grotesque sort of way, he was really rather handsome. There are two subspecies of orangutan, one living on the island of Sumatra and the other in the jungles of Borneo. Adam was a Sumatran orang, and at twelve years old he was in his prime. He was a massive beast, covered almost completely in a long auburn matting of fur. His small, slanting eyes, huge jowl-like cheek pouches, and long gray beard and mustache made him look for all the world like an aging Manchu emperor or, perhaps the high lama of Shangri-la. Adding to this oriental appearance was Adam's red fur, which mantled his broad back and hung in long strands down his arms like the wide sleeves of a kimono.

Adam was the introvert of the family; he was exceptionally placid and rarely moved from one corner of the cage. Here, on a large pile of straw, he sat in yogilike contemplation for most of his waking hours. Occasionally, if he felt the scrutiny of the visiting public too keenly, Adam would pick up a handful of straw and, with infinite slowness, place it over his head, so blocking out all wordly distractions. After two or three minutes like this, he would carefully move his straw shield to one side and peer out to see whether the source of his distraction had disappeared. If the visitor continued to stare, Adam would turn away and, imitating the Buddha, continue his meditations facing a wall of stone.

The female of the pair, Mandy, was a very different sack of potatoes. She was a member of the Bornean race and, like Adam, had been taken from her native jungle at about two years of age and shipped to England. Whereas Adam, even in captivity, retained much of the dignity of his species, Mandy seemed to be nothing more than a large mound of lard covered in straggling purple-colored hair. She was unbelievably obese and resembled nothing so much as a fat eighteenth-century fishwife. And she had the manners to match! Not for her Adam's aloof solitude. Mandy seemed to want to get her own back on all those ignorant humans who spent their time standing and gawking at her.

She liked nothing better than sitting up against the bars of her cage, resting her huge bulk on her elbows and looking uncannily like some elderly village gossip, while her gimlet eyes scanned the crowd for a likely victim. Once she had selected her prey Mandy reached out her left hand as if in supplication, while at the same time her right hand, out of sight behind the broad beam of her hips, gathered up as many pieces of rubble as she could get hold of. Usually the unsuspecting visitor would lean forward and try to touch Mandy's hand. Then, with uncharacteristic alacrity, Mandy would strike, showering the luckless spectator with stone and sand.

The strange thing was that Mandy seemed to get no pleasure at all from a successful fooling of the enemy. Whereas the chimps in the next cage, who also used this ploy, used to whoop and scream and stamp their feet in triumph whenever their shots hit their targets, Mandy's face and whole demeanor remained impassive. The most she would do would be to try to get in a second quick strike before the visitor retreated. When she got bored with this method of attack, Mandy would draw the visitors even closer to the bars and then spit in their faces. She was a very good shot.

Looking at Mandy it was easy to understand why the orangutan has remained the poor relation of both the chimpanzee and the gorilla as far as psychological research is concerned. In adulthood the caged orangutan is the least prepossessing of the great apes. The species seems little more than inert, rufous-furred lumps, only roused to ponderous activity by the appearance of the food tray. Chimpanzees, by comparison, are lively and vivacious until well into old age, and even the gorilla radiates more potential for movement than poor old *Pongo pygmaeus*. The impression of full inactivity in zoo orangs is overwhelming, and it is small wonder that the consensus among psychologists is that to study the orangutan is at best rather like working with a dumb chimpanzee.

This opinion is highly suspect. While the orangutan is probably more retiring than the other two great apes, it is also true that zoos make no allowance for the orangutan's tree-climbing way of life. In their natural environments, both chimp and gorilla spend most of their time on the ground, whereas the orang lives almost entirely in the high canopy of the rain forest. Yet captive orangs are often deposited in bare cages containing very little climbing equipment. The orang, evolved over millions of years for nothing other than climbing, is the trapeze artiste of the great apes, and anyone who has seen this gorgeous creature at home in the trees will realize why, in a bare prison cell, it loses most of its interest in life.

And then we take this imposed lethargy as evidence of the creature's indolence and unintelligence! It is rather like placing a St. Bernard in a tree and then saying "all dogs are lazy" because it does not move! I was sure that if an infant orangutan were raised in a stimulating environment with plenty of opportunity to exercise and climb it would display a much higher level of intelligence than had previously been thought.

Adam and Mandy were fascinating, but by now I was very impatient. What I really wanted to see was the five-week-old baby ape that had monopolized my thoughts since the zoo had telephoned to say my study was approved. At last, the zoo curator, John Laing, suggested we move on, and we continued past the chimphouse toward a building at the far end of the ape enclosure.

"This is the place," John said. I looked up to see a wooden building about 20 feet long, ten feet wide, and perhaps eight feet high. It was divided into two rooms, one of which was completely visible, the nearmost wall being all glass, sloping outward from base to roof. The other room had a plain wooden door and only one small window through which I could discern nothing at all about the interior. John dug deep into the voluminous pocket of his windbreaker, extracted an enormous set of at least forty keys, and with the practiced ease of an expert selected the right one for the job. The door swung open, and in we went.

The room was very gloomy, and on the air there was that musty, almost rancid odor that I had smelled with Lucy and that I now always associate with the great apes. Standing in the doorway I could dimly discern a sink to my right and, directly in front, about five feet away, two large wooden cages, one on top of the other. At the base of these was a small portable baby bed, almost hidden against the background of dark brown floorboards. I stepped forward, to see what lay in it.

Immediately a weird humming noise filled the air. Hurriedly leaping back, I looked around, certain I had sprung some newfangled alarm system. John's face remained impassive. "Thermostatic heaters—blow warm air around the room—keep it about seventy-five degrees all the time." I moved toward the crib again.

I don't know what I expected to see. I had never seen a baby orangutan before, but never in my wildest dreams could I have conjured up the pathetic little huddle before me. There, lying face down on a blanket, was a tiny splodge of auburn hair, not one foot in length. Through the hair one could only just distinguish a head and body. From this hirsute mass projected an upper and lower pair of scrawny limbs, the top pair seemingly twice as long as they should have been, all four limbs ending in pitifully small hands that clung desperately to the soft covering of the blanket.

John squatted down beside the bed, and I did likewise. "There he is," he said. "There isn't much of him," and here a hint of pride entered his voice, "but he'll do."

As if to reinforce this statement of his adequacy, the brown splodge suddenly jerked and began to cry softly, his head making urgent side-to-side movements obviously designed to put the infant in contact with the breast and milk. We watched him for some time, intrigued by the instinctive search pattern, as basic to an ape as to a human. But what I found really fascinating was the baby's ear. It was the only patch of pink on an otherwise brown body, a tiny pointed whorl that looked as fragile as porcelain and so like my idea of an elfin ear I could hardly believe it was real. I was still staring when John stood up.

"Ethel looks after him, she's a keeper in the ape section." He looked at his watch. "It's one o'clock. You'll be able to see her in three hours' time if you come over to the animal kitchen—keeper's tea break." He opened the door to leave. "By the way, I call him Cody, short for Kodak!" John was half smiling.

"Kodak?" I echoed, genuinely puzzled.

"Yes," John said, and for the first time I saw a broad grin on the curator's face. "He's a little brownie."

After John had left I pulled out my notebook, scribbled "CODY DIARY, Observation I" on its first page, and sat back to begin my first observation period with the infant ape. Ninety minutes later I had made my first discovery: at five weeks of age baby orangs, like baby humans, do nothing but sleep. Except for a jerk or two, the past hour and a half had been as uneventful as if the cot had been empty. I was sadly disappointed with his lack of activity, but the interlude did give me time to reflect on the way things had gone since I had left hospital.

The more I thought about it, the more I realized how lucky I had been in finding Cody. There are no more than one or two orang births in Britain each year, sometimes none at all, and fate had been enormously kind in allowing the end of my recuperation to coincide so satisfactorily with one of these rare births. This timely happenstance would allow me to chart the infant's development over almost the whole of his first year of life, perhaps even longer. Also fortunate was the curator's information that records of weight, measurement, and feeding were available for the few weeks that had intervened between the infant's birth and the beginning of my study.

But not everything on the horizon was so rosy. The only accommodation that the zoo had to offer was a dilapidated ten-foot trailer of uncertain vintage, with no running water, no heating, and no loo. It was hardly the ideal lodging place for a young man whose doctor had advised one full year of "easy living" after the beating my liver had taken with hepatitis. Worse still, no one at the zoo yet knew of my plans to teach the young ape to talk. I was so afraid of being thought a little less than sane—and being denied a suitable subject for my research—that I had thought it best to mention only the more innocuous parts of my study. The upshot was that the zoo had accepted me solely as a student of the baby ape's physical development—the growth of his limbs, his increase in weight, his time of teething, and so on. I had no idea how I was to progress from this point to an acceptance of talking tuition.

A slight movement and a plaintive cry shook me from my reverie. Looking up I saw that the baby orang was no longer asleep and that he obviously believed it to be far past his dinner time. He tried the blanket several times with his lips, rejecting the nonproductive material after a few sucks and writhing in frustration with such violence I thought he was in imminent danger of upsetting his bed.

On one occasion as he turned his head his lips accidentally struck his own hand. At this stage, I knew, the human infant cannot distinguish its own hand from anyone else's; so I watched Cody's reaction with interest. Sure enough, the little ape fastened on to his own fingers as if they were a certain source of milk. Only after a 30-second suck did he decide that his lips would be better occupied elsewhere. Then the squirming began again, and his cries became angry and demanding.

I felt as if I should do something to aid this helpless bundle of frustrated fur; so I tried talking softly to the infant in typical baby fashion. "There theres" and "coochy coos" seemed to quiet him slightly. The next step was naturally to pick him up.

For the first time I could see his face clearly, and it was not a pretty sight; only a mother could have loved that ugly mug. His face was so squashed and creased as a raisin, with wrinkled bags under and almost obscuring his eyes like those of an old man. His hair added to the overall picture of premature senescence; for not one tuft graced his scalp from eyebrow to crown. As though self-conscious of this fact, the rest of his nut-brown tresses were brushed forward "Nero-style," craftily giving the impression of a well-endowed cranium. The first thing I noted was that Kodak was very light, not more than seven pounds in weight and definitely lighter than most human infants at five weeks of age. Considering the size of Adam and Mandy, I found it hard to believe that this tiny mite could ever grow to even remotely resemble his parents.

It didn't seem possible that he could eat enough to put on so much bulk. For his part, the young ape seemed oblivious to everything except food. His eyes stared vacantly toward the ceiling, and he had great difficulty in getting both to move in the same direction at the same time. As a result, they frequently wandered off on their own, leaving the poor infant cross-eyed. Only his mouth seemed to be in perfect working order, and it was obviously straining for something more substantial to work on than his nonproductive blanket.

As soon as he had been picked up, his head had started its side-to-side movement, sampling my coat, its buttons, and everything else his mouth happened to light upon. He was definitely very hungry, and I felt quite annoyed that no keeper was at hand to feed him. As I looked around for help my gaze fell upon a baby bottle full of milk. It looked like a pre-made meal—and just what Cody needed. It was not really part of my job to feed the young ape, but he seemed so hungry I carried him over to the workbench and reheated the bottle to what I thought was a suitable temperature.

As soon as the bottle touched his cheek Cody turned his head and fastened his mouth avidly on the teat. His little mouth was just one big pout, and his cheeks vanished, replaced by dark hollows as, with eyes tightly shut, he sucked at the sweet milk with all his might. I was astounded: my sister's child, Louise, managed two or three tiny sips of her milk and then rejected the bottle for a while, but this little chap just took hold and never left off until the last drag was consumed. Even then I had to prise his lips from the bottle. Delighted and moved by his actions I bent forward to stroke his head.

Immediately the little ape jerked so violently that he almost threw himself out of my lap. I had forgotten that very young infants of any species have not yet learned what things can safely be ignored and what objects and actions must be attended to. As a result, they at first overreact to all unusual stimuli; and with their very limited experience, almost any happening is unusual. Adults act similarly when placed in unknown situations. My first attempt at driving a car was a nightmare—every pedestrian seemed about to leap under my wheels, and I often braked unnecessarily at cars that I mistakenly believed to be turning, slowing, or accelerating. Until one learns to "read" the road, one is in a very similar world to the newborn infant. Simply bringing my hand toward Cody's head precipitated his violent response. It was another piece of datum for the notebook—"startle reflex very good."

I was grateful to the young ape in another way too. Cody had unwittingly reminded me that there was one set of tests I must give before proceeding with any others: the infant would have to be tested for the various reflexes. His startle reflex I already knew was good, but I needed to test grasping, Moro, Babinsky, pupillary, and blink reflexes before I knew I could take Cody to be what he seemed—a healthy member of his species.

These are the same reflexes that a pediatrician uses when testing human infants, and the reason for such an examination is the same if the child is human or pongid. They give the first indication of whether or not the child is suffering from congenital neurological damage or whether it has suffered similar trauma during or after birth. This was especially important to me as it would not really be of scientific interest to chart the development of a mentally subnormal orangutan when there was still nothing known of how a healthy orang infant grew and developed. In fact, "Is he normal?" had been one of the first questions I had asked the curator on being told that I might study the young ape.

I knew that Kodak had been removed from his mother and was being hand-reared away from her. I was worried that this meant he was in some way incapable of doing everything a normal newborn infant should do. John Laing, however, was able to reassure me that, as far as he knew, there was nothing unusual about the baby orang. Mandy had begun labor at around twelve noon, and within an hour and a half the baby was born. This was pleasing information; long, difficult births can often result in brain damage, usually because the blood supply to the infant is somehow curtailed.

But while the baby seemed to be all right, his mother had appeared quite distraught. She did not seem to know what to do with her infant, and while she did not mistreat him, neither did she allow him to feed. Mandy made no attempt to place the infant to her breast to nurse, and each time the baby fortuitously found her nipple he was unceremoniously pushed away. Kodak was even balanced on his mother's head for a while! Finally, after the baby had gone for more than 15 hours without food, John Laing decided reluctantly to anesthetize the female and remove him before he died of starvation.

Since that time the little orang had spent his time in the small hut at the end of the ape enclosure. It was a rather impoverished environment, but I planned to change all that. I would soon transform his dull surroundings into something very similar to a human infant's playroom. Then we would see just how far Kodak could progress.

But at first, I had to be sure my "subject" was quite normal. I decided to test his grasp reflex. A pencil placed across his open palm produced an immediate response. Without looking at the object or in any way noticing its existence, the little ape closed his hand smartly and tightly around the pencil. "Reflex grasp present," I scribbled in my notebook. Human babies also have this instinctive response, but it is one that is far more important to an infant orangutan. Whereas human babies can grasp strongly and can even support their own weight, they have little need for such an ability. By contrast, an orang infant must be capable of clinging to its mother's fur almost from the instant of birth. Without a strong grasp reflex, the infant is almost certainly doomed; it will fall from its mother and crash to its death through the canopy of the rain forest.

So Kodak was all right on that count. Now the Moro reflex. Even as I began to move the ape, I noticed a strange expression on his face. Intrigued, I sat him up and watched as he pulled his mouth so that his lower lip vanished under his upper lip, which was itself extended into a sort of vase-shaped pout. His eyes too looked almost painfully strained. Then, as suddenly as it had begun, Kodak's stressful expression began to relax, and immediately I felt that hot, wet sensation that told me I had been peed on! Before I could even begin to get angry, the young ape closed his eyes and went to sleep.

5

For the next two months I was so hard at work, I had no time to worry about the "talking" problem. Every day, from first light to well past sunset, was fully taken up with learning to be an unmarried father. It was not the most pleasant period of my life and my experiences came near to putting me off parenthood forever. During the first month, Cody required three-hourly feeds of baby milk throughout the day and night. A half-hour walk from my trailer to the infant's cage at three in the morning did little for my health and even less for my morale. I had become a nursing note-taking automaton. Making Cody's 3 A.M. bottle and feeding the young ape was a mechanical routine although I never quite got used to changing a soiled diaper at this ungodly hour. When this nightly ordeal came to an end, my joy knew no bounds. And when, two weeks later, I was advised to feed Cody only three times a day—at 8:30 A.M.,1:30 P.M. and 5:00 P.M.—I found the work much easier and could actually take some pleasure of my "pupil's" progress.

After several weeks of doing very little except sleep, Cody finally decided that the conscious world might be more interesting. During my strict observation periods I used a shorthand code to keep a record of his every movement. I also noted any advance that occurred during the other parts of the day. I now had a complete record of how, over the past few weeks, the young ape had developed from almost total inactivity into a lively, though still immobile, ball of vibrant energy. Cody was no longer a mass of simple reflex movements, but an increasingly coordinated and independent creature. There would be no time in his life when he would see so many new things, or progress quite so rapidly; and it was fascinating, each day, to observe the way in which every aspect of his being—vision, hearing, touch, taste and movement—increasingly came together into a coordinated whole.

One example of this was Cody's reaction to sound. In all primates, the adult of a species uses the sounds he hears to help him to construct his world and to live more safely in this world.

We humans stand at a roadside bend and listen to the sounds of approaching cars, squirrel monkeys hear a noise in the branches and flee from a stalking anaconda, and orangutans use sounds to locate one another in the dense Asian rain forest. Sounds are not simply noises: we connect the sound with an object or with some sort of activity—a clap of thunder means an approaching storm. Usually, if the sound is nearby, we can locate its position, and if it is interesting, we turn our heads to see what made the noise. For a newborn infant, it is all quite different. Baby orangs (and baby humans) do none of this; they have yet to make the association between sound and object, and the most they will do is to "start" at an unusual noise. Somehow they develop the object-sound ability although, at the time of my research, no one was quite sure how this happened. By living with Cody for most of his waking hours, I was able to follow the process minutely and to discover the more important stages in the infant's progress to "looking in the direction of an unusual noise."

What I did was to stand directly behind the infant while he was absorbed in some activity such as sucking his blanket. I then moved to one side, taking care not to be seen, and made an unusual noise; sometimes it was a "cricket" clicker, sometimes a burst of music or a yell, but always a loud noise that Cody had not heard for some time. It was a simple method, but the infant's responses were illuminating. Up to the seventh week, Cody's reaction to a noise, from no matter what source, was simply to jump almost out of his skin. During the eighth week, he no longer started but instead ceased all activity as soon as he heard a strange sound. Cody was beginning to pay attention to the noise, and I was interested to note that, in this same week, he also became very still and quiet when an object was passed in front of his face, or when I placed an object in the palm of his hand. These were the first signs of the infant "tuning in" to the outside world, attending to stimuli that were novel and therefore of interest to his young mind.

The next step followed very soon after, in the ninth week, and it was a response that I had never seen reported in scientific papers on the subject. Whenever I made a sound, Cody would rise from his prone position so that he was up on all fours; then he would look all around him before sinking back into his blanket. Within seven days of this response, Cody began to show definite signs of looking in the correct direction of a sound produced on one side, and by his tenth week he could locate an unusual sound as well as any adult. He had achieved mastery of this skill at approximately the same time as a human infant.

At this point, I ended my tests and tried to analyze the results. What really intrigued and puzzled me was Cody's "head up and look all around" response. Why should he look about him when it seemed far easier to look only to one side? Then I realized what must have been happening. Some time before the infant knows how to localize a sound, it seems he learns that sounds often presage interesting events: the tinkle of glass and the sound of boiling water precede feeding, the noise of a lawn mower is audible before the mower heaves spectacularly into sight, and so on. So, whenever the infant hears a sound, he first looks about him in the hope of spotting the source. It is only later that he "realizes" that the ear on which the sound waves strike first is the side to which the head must be turned in order more quickly to espy the interesting sight.

What it comes down to is strategems: the child uses the strategem that gives it the greatest return for a certain amount of effort. When the child did not look up at an interesting sound, it saw little; when it did, it saw more; and when it turned its head to the side from which the sound appeared to emanate, even more. We all live by these strategems or rules, but until we can see them in so simplified a form, it is hard to recognize the rules which govern most of our daily lives.

Cody was rarely quiet for long, and I was amazed at the way his restless activity slowly but surely increased his range of discrete movements. It seemed as if a new muscle group was exercised and added to his repertoire every two days or so. Just as an athlete gradually strengthens and improves himself for the big day, so Cody would train diligently for his own personal ambition, which during this period seemed limited to gaining at least partial control of his body and wayward limbs.

His first achievement had been, on my third day at the zoo, to push up his head and chest so that he rested on his elbows and forearms. After this, there were periods when he raised one side of his body completely off the ground, and pushed both legs into the blanket and raised his hips (almost asphyxiating himself in the process). Then, while doing his up-on-elbows trick, he would lean rapidly from one side to the other, or push his legs far down the bed only to retract them immediately or perhaps raise them in the air. Gradually he coordinated each movement with every other so that they formed an integrated whole. The look of concentration on that tiny brown face was at times excruciatingly funny, but to Cody it was obviously no laughing matter. He appeared intent on leaving the bed as soon as possible, probably as a consequence of his burgeoning interest in the objects of the outside world, especially people objects.

Cody was fascinated with people: unlike other objects, they actually moved! I was a prime target for his scrutiny, and as far as his limited ability would allow, Cody's soulful eyes would follow me about the room as I prepared his meals. It may have been filial devotion, but I had the sneaking suspicion that I was construed as a giant food bottle rather than a person!

By the time he was three months old Cody had come a long way from the inert ball of fur I had first seen. Unfortunately, the list of his nonaccomplishments was still embarrassingly formidable. Cody was not developing as fast as I had expected. A chimpanzee will show signs of crawling at around three and a half months, but Cody demonstrated no such inclination. In fact, I was to wait a further four months before the young orang even attempted this method of locomotion. At three months, he was practically incapable of moving his whole body; he had just learned to twist around on his belly so that his head faced the bottom of the cot. He was unable to grasp on sight, or stand without support, or even to control his hands.

He could not sit to save his life—it was to take him 26 weeks to achieve this skill (a human baby takes about two weeks longer). At 12 weeks, whenever I sat him upright he would collapse forward, his hands at his sides, making no attempt to save himself from the inevitable bump on the floor. Time after time I tried it, and time after time Cody would end up with his nose against the boards, staring in shocked silence at the floor. Luckily it did not seem to hurt him.

It was as if he did not recognize that his hands were his own. Until he was more than two months old Cody would spend long periods of time gazing at his fingers, fascinated by their wriggling and obviously practicing his control over their movements. But if ever, during one of his arm-flailing sessions, Cody's hands passed close to his face, he would utter a little squeak of mingled fear and surprise, his tiny body jerking back with fright.

It was only around the eighth week that Cody began to gain some control over his wayward digits. Then his main occupation was to suck them: one finger, two fingers, the whole hand, every variation on hand-sucking was tried out—even sucking two while at the same time the rest beat out a tattoo on the end of his nose. But as any baby will tell you, at this age it is very difficult to get your hand directly into your mouth. To bring the reluctant limb to the face is child's play, but you are just as likely to miss your mouth as hit it, and several times I thought Cody had poked out an eye or was picking his nose in his wild attempts at hand-to-mouth control. On one occasion, he even managed to get his finger in his left ear, his half-open mouth sucking expectantly on the empty air and a puzzled expression clouding his eyes. Where *was* that hand?

Cody finally learned hand-sucking at around ten weeks. Even after he had accomplished this trick I could not tell if he really knew his hands were his own, or whether he thought they were like his other toys, just something to look at or chew on. Apparently it was the latter, at least until the eleventh week. On day four of this week I saw Cody suddenly stop and stare at his arm. He looked first at his hand and then slowly moved his gaze up his arm. With mounting horror and surprise his wide-eyed stare moved from wrist to elbow and then up and on to his shoulder. The skin on his forehead shot back, a sure sign that he was disturbed. "Good Lord!" he seemed to be thinking, "that object's attached to me!" It took him several days to regain his composure.

Despite the long hours I spent with Cody and all the petting, stroking, comfort, and feeding I gave him, for a long time the little ape gave me no thanks at all for my pains. This was rather discouraging. I had hoped that a bond of affection would grow up between us so that if and when I got the chance to train him in "talking" we could work together as much more of a team. For my part, the bond was already forged; two months spent with such a beautiful little creature was more than enough to make me very fond of him. But on Cody's side there was nary a glimmer of recognition.

Then, during his fifth month, suddenly he realized that people were different; and I, presumably because I had spent most time with him, became singled out as "mama." This behavior corresponds quite closely with human children in that, as any mother can attest, there is a point when her child suddenly begins to give her special attention.

Although very pleased and honored, I soon discovered that this distinction had an important drawback. While previously I had been able to put Cody down in his cot after feeding, bathing, etc., and walk quietly away, now my every departure was certain to bring forth the most violent and voluble remonstrations. Each hand (and practically speaking, Cody had four of them) had to be prised from my shirt; and immediately I turned my attention to removing another clinging limb the first one reapplied itself with renewed vigor. When I finally freed myself from this furry octopus, he at once exploded into vocal disapproval, and for such a little chap his voice was positively stentorian—a long piercing scream punctuated by a rasping grunt, as he took in another lungful of air in preparation for the next mind-splitting blast.

Cody's feeding habits were very similar to those of human children of his age. Beginning with half cream at birth, he soon graduated to full cream at two months. At three months I added pre-cooked cereal, "pablum," to his diet, and at around six months of age I planned to supplement this regime with canned baby foods. When I did eventually try canned foods, it was under somewhat trying circumstances. I had ordered these provisions over the phone from a local pharmacy. Unfortunately, owing to some logistical malfunction, when the cans arrived they were all labeled "pork stew." Orangs are not known to eat meat in the wild, but I decided to let Cody make up his own mind.

When mealtime came he opened his mouth eagerly, fully expecting his usual porridge and honey. The spoon, full of pork stew, went in, and his jaws clamped tightly around it. I withdrew the spoon, but instead of the usual hurried swallow and eagerly reopened mouth, Cody's face suddenly went quite rigid, his eyes staring straight ahead. I could almost hear him thinking, "What is this stuff?" His eyes swiveled upward in their sockets, and he glared at me suspiciously, but his jaws steadfastly refused to work. He would not swallow the stew; but as it was definitely food, and as food was his *raison d'être,* neither would he spit it out. I was seized by an almost irresistible desire to giggle at the silent pantomime being acted out on my knee.

Frowning hard, Cody looked distrustfully from his dubious

breakfast to my face, to the breakfast and back to my face, seemingly trying to decide whether this was an attempt on his life by poisoning. I must have an honest face, for after about a minute or more the jaws began moving, slowly at first, then more rapidly. The second teaspoon got the same treatment, but the delay between taking the spoon into his mouth and the final chewing was slightly shorter. It was shorter still on the third and fourth spoons, and by the fifth Cody was wolfing down this new addition to his menu as if it were his favorite dish.

Solid foods—bananas, oranges, and boiled eggs—were added to Cody's diet a month before his first birthday; and while each was subjected to a period of testing (Cody was not too keen on oranges) none was ever rejected. Unlike most parents who worry over their offspring's refusal to eat, my problem was knowing when to *end* a meal. At mealtimes Cody was little more than a mouth machine, and had I allowed him he would have consumed far in excess of his needs. This seems to be a characteristic of the species; almost every zoo orang is terribly obese, primarily because they are fed too much, but also because the orangutan finds it impossible to say no. It was only during a bout of gastroenteritis that Cody refused his food, and then I realized just how anxious human parents must become.

At one stage, I was so concerned over his refusal to drink that I attempted to "milk" the teat while it was still in his mouth. Unfortunately, in my eagerness, I pulled a little too hard. The teat came out, and I suddenly found an albino-faced orang on my lap. Cody endured this unusual shower with his usual stoicism, not uttering a sound, his mouth set firm and his eyes glowering in mute rebuke. This silent indignation was really disconcerting. If he had only cried, I could at least have told him to "be quiet it's only a bit of milk." But this cutting reproachful stare left me no alternative but apology, which he also greeted with haughty disdain.

Apart from gastroenteritis and the danger of pulmonary infections, both of which were a constant threat to Cody's health, I had found my first three months of parenthood less of an ordeal than I had expected. True, Cody did have a bad attack of diaper rash, but I was coping with this by liberal applications of petroleum jelly to his nether regions.

So I was coping with Cody: in fact my main problem was not the ape, but a human. Ethel, the keeper who had looked after Cody when he was first born, was quite jealous of the time I spent with "her" baby. She took every opportunity to keep Cody away from me, and to show me up as an incompetent "keeper." When she heard I was thinking of teaching Cody to talk (I had broached the subject with the zoo keepers, not the management), she blew her top and stormed into Cody's nursery demanding that I give up any such "mad ideas." I was washing soiled diapers at the time, so I was not in the best of humor. We had a vituperative argument, after which we both stomped out in a fine old rage. By chance, while still in the full flood of my anger, I bumped into the zoo director, Mr. Jackson, out on his daily rounds.

The director was a large rotund man, whose avuncular features concealed an acute business mind. He asked me how I was, and before I knew it I was pouring out all my frustrations and ideas, including the "talking" project. Like the astute businessman he was, Mr. Jackson ignored most of this diatribe. His sharp mind concentrated on one thing: commercial possibilities of a talking ape.

I tried to be just as sharp and poured out a rather garbled history of ape language, glossing over the dearth of sounds learned, and impressing on him the fact that sounds and words had been taught to apes, that orangs seemed to be the best "talkers" and that I was sure I could train the zoo's infant orang to talk to his satisfaction. I paused for breath. The idea was out at last. Would it be yea or nay?

Mr. Jackson began chuckling. "My God," he said, "what you scientists will think up next I don't know. I've never heard the like. Do you think you can do it?"

This was no time for wavering. "Certainly," I replied, trying to give the impression of confidence. "There's every chance."

Mr. Jackson looked at the bears in the cages to the right, not seeing them. He looked back at me. "Very well then. I'll see John

Laing about it. I don't think it's possible, but even if you don't get a peep out of him, the publicity we get will be worth it. When do you think he'll manage his first word?"

I was so taken aback by this sudden acceptance that for a few seconds my voice refused to operate. I finally managed to gasp out that around ten to twelve months might see the infant producing a sound approaching a human word.

"OK, then, we'll see how it goes," he replied and turned back toward his office, slowly shaking his head. I could still hear him laughing at the stupidity of the idea, but it didn't matter. I looked around at the sky. Where was the thunder and lightning? Shouldn't the earth be shaking? They were going to let me try!

6

Cody was not the only ape at the zoo who interested me. Mandy and Adam, his parents, also made fascinating subjects. Just as in human groups, the marked differences in temperament between the two adult orangs reflected different aptitudes and abilities. I was intrigued by these discrepancies and decided to look in detail at one specific "talent"—the orangutan's artistic ability. Not only could I use the information from Mandy and Adam as a comparison with Cody's own skill, but the data would also be the first of its kind obtained for the adult of this species. At that time only the chimpanzee had been given the opportunity to draw.

If a chimpanzee is presented with a sheet of blank paper and a pencil, it will very soon learn to scribble. This in itself is quite an achievement—very few monkeys (the South American Capuchin is one) show even the slightest interest in "taking a line for a walk." But there is more to it than simple scratch marks on paper. The chimpanzee will usually concentrate its scribbling in the *central portion* of the page as if it considers this area a separate "drawing space" and wishes to draw attention to it.

That the chimp understands the idea of paper as drawing space is shown by another experiment. Here, a one-inch margin is drawn around the paper before it is given to the "artist." The chimp will then usually scribble only within the area enclosed by the margin, taking great care not to cross the line. If the paper is marked with a black circle in the middle of the left side and midway between the top and bottom, the ape will often try to "balance" the lopsided appearance of the page with a scribbled circle drawn in a mirror image position on the right-hand side.

Chimpanzees do not only express themselves in monochrome: some have become accomplished painters in oil or water colors. Color was something my subjects *might* progress to, but as a first stage I intended to try simple pencil drawings.

Three forty-five on a Saturday afternoon saw Liz (my finacée) and myself on our way to the apes' quarters. Our appearance was responsible for quite a few raised eyebrows among the scattered visitors who had braved the biting winds of a Yorkshire January to sample the delights of our menagerie. Liz wore a twelve-inch wooden frame slung over her head and around her neck, like some outsize timber necklace, and staggered along under the weight of at least a ream of drawing paper. We probably didn't need that much, but I have always been the careful, just-in-case sort of researcher. For myself, I carried what to untutored eyes may have looked like a two-foot square of wood glued on to an eight-foot plank, but was in reality a carefully thought out and skillfully constructed piece of scientific apparatus designed to chart the very limits of anthropoid esthetic capacities.

The plan was to show the orangs the paper and pencil and to demonstrate to them the action of drawing. Then we would fasten a sheet of paper to the blade end of our giant shovel using Sellotape, after which we would slide it under the cage door, where there was a three-inch gap. We would then give either Adam or Mandy the pencil and with a bit of luck, one of them would draw on the paper. If it worked, we had decided that as soon as there was a definite pause in the doodling, we would remove the drawing by pulling on the shaft of the apparatus, thereby bringing the paper to the outside of the cage. Here we could remove the drawing for analysis, substitute a fresh piece of paper, and return it for a second period of artwork.

Now the adult orang is not renowned for its exuberant interest in anything, let alone so esoteric a subject as art; so our confidence in even a limited success was hovering around the zero level as we made ready to slide the board under their cage for the first time. Adam decided straightaway that we were not food givers and that we were, therefore, beneath contempt. He took himself off to the high shelf of the cage and studiously ignored us. Mandy, however, came lumbering across to our side of the room. Grabbing two handfuls of straw to cushion her elbows on the cold concrete sill, she settled herself directly in front of me.

"This is a good sign," I said to Liz, at the same time bending to pick up pencil and paper for the drawing demonstration.

It was lucky I ducked when I did, for had I not, the well-aimed gobbet of saliva that splashed with considerable force against the wall behind me would have hit me right in the face. I turned to see Mandy marshaling another mouthful of spittle in preparation for a second try.

Liz fled. I was tempted to follow as well, but the thought of all those hours spend preparing the drawing board the night before, all that screwing, sawing, and hammering, made me determined not to let her beat me. I pulled up the hood of my anorak and pressed on, casting a friendly smile in Mandy's direction, and all the while scribbling madly on paper already wet with the spray. Gobbet after gobbet rained against my hood. I was sorting my fourth sheet of paper when I suddenly noticed that the storm had subsided.

Cautiously, I peered around the side of my hood. Wonder of wonders! Mandy was sitting absolutely still and staring intently at the pencil and paper on my lap. She seemed to be taking an interest at last. My confidence renewed slightly, I smiled at her, but received a blank stare in response. Then she suddenly pushed her hand through the bars and held it out appealingly, her eyes staring fixedly at the pencil. Once again I scribbled away and could see the piglike eyes following the pencil. Perhaps my patience was paying off, and it would work after all! She was still begging. Quickly I pressed the pencil into her great hairy paw. Just as quickly, Mandy transferred the pencil from hand to mouth. The massive jaws moved, and helpless to prevent it, I could see my lovely pencil being slowly ground to pieces between her enormous yellow teeth.

Out came the hand again. I had one pencil left, and much against my better judgment I handed it over. She spat out the remains of both pencils twenty seconds later—a pale mash of broken wood. Then, these interesting objects having proved inedible, she returned once again to the more entertaining business of spitting at me. Containing as they did a few pencil chips, her missiles were a little more painful now.

I had had enough. I spat back. It was a good shot hitting her plumb on the end of her nose. She didn't like it at all and made for the far side of the cage where, a great heap of red fur, she sat glowering at me from behind a pile of straw. I retrieved the art board and, my dignity somewhat restored, left the cage.

I was despondent: we had nothing to show for all our efforts. It seemed that the adult orang was less interested in being creative than in destroying anything it could lay its hands on. Perhaps such behavior was a result of zoo life. I could only hope that Cody would prove more amenable to my study methods.

7

The Tannoy boomed out its festive message to an all but deserted zoo: "We wish you a Merry Christmas, and a Happy Nooo Yeear."

To be honest, as I prepared yet another spoonful of food for my fosterling, I did not feel too replete with Christmas spirit. For one thing, Christmas had been over for three weeks; and for another, the project was not going very well. Cody was now nearly six months old, and it had been almost four weeks since I had begun what, for me, was the most important part of my work—trying to teach him to talk. Before that his lack of psychological awareness had prevented me from attempting talking tuition, as I had rather grandly termed my training methods. But so far, despite many hours of patient tuition, Cody's linguistic education had been an unmitigated disaster.

For lack of a better method I started by teaching the little ape with the same techniques as Catherine and Keith Hayes had used with their chimp, Viki. The Hayes had not achieved very much with their chimpanzee during the six years of training, but I was pinning all my hopes on the orangutan proving a much brighter pupil as far as talking was concerned. The Hayes had used a shaping method of conditions to produce Viki's few sounds: they rewarded her (with milk or food) for any sound she made. Then, when she had learned to produce a sound on command, the psychologists gradually tried to *shape* her vocalizations toward a target word by rewarding only closer and closer approximations to the sound they wanted. When necessary, they also indulged in a little lip manipulation to produce the desired result.

Although this technique had not been too successful with the chimpanzee species, I had hoped Cody would turn out to be more tractable. But the truth was that the shaping method seemed to be of even less use with an orangutan. To begin with, unlike the chimp, Cody very rarely made any sound, so there was nothing to reward! What I needed was a way to persuade Cody to emit some sort of sound. As a vain hope, I asked the advice of the two dolphin trainers at the zoo, Beth and Paula. I had seen their show a few days previously and been very impressed by the tricks they had taught their charges. In addition to the normal leaping and ballthrowing, Beth had trained three of the dolphins to "sing." While she waved her arms over the pool in imitation of an orchestra conductor, the cetacean trio would rise from the water and give voice to a wavering, high-pitched song. Dolphins are not the most vocal creatures out of their natural element, and I wondered if Beth's training methods might work for Cody.

At first Beth was reluctant to divulge the secrets of her training regime. But after a solemn promise that I would not reveal the method for at least two years, she at last agreed to help. She suggested that I offer Cody his favorite food and, just as he opened his mouth to take the spoonful, quickly pull it out of his reach. He would scream out his frustration, and then I could reward him for the vocalization. Gradually, Cody would (or should) come to relate screaming with food. This method, I was told, works well with dolphins as a first step toward teaching them to sing. After that, Beth told me, I was on my own: she had given away far too many secrets already. I assured her it was all I needed: once Cody was vocalizing I felt sure I could shape his sound into a reasonable approximation of some English word. To me, Beth's method seemed to hold the promise of success, and I lost no time in trying it out on Cody.

Orangutans are not dolphins. It was not that Cody learned nothing from this experiment. But what he did learn (after only two or three mealtimes of screeching indignantly each time his food disappeared) was not that "scream equals food" but that if he did not scream and just sat silently, in a very little while the food would return. He seemed to know I was so afraid of him starving I would always make sure he got enough in his belly. Being in sole charge of an animal worth over $7,500 on the open market was an enormous responsibility, and I could not take chances with his health, no matter how much I wanted him to speak. The university had not insured me against the loss of Cody, and I simply could not find that sort of cash should something go wrong while I was caring for him. And besides, like all good parents, I was happiest when he had a substantial meal inside him.

So for the twenty-eighth time that meal, as I swiftly whipped a spoonful of cereal away from Cody's open mouth, the hairy midget simply clamped it shut and sat immobile on my knee, smug in the knowledge that his food would very soon return. I had become so desperate that I even tried dangling him by his arms in the hope that he would cry out in fear; but all this had accomplished was a little pool on the floor. Or worse. "Speak to me, Cody," I begged, "please, just a peep." Nothing. Not so much as a whisper or sound.

I stood up and despondently reached for his baby bottle. Immediately, Cody gave a little cry of fear. For a moment I ignored him. Then it hit me—Eureka! I had been meaning to pick Cody up and feed him, but as soon as I realized the significance of his sound, I quickly squatted beside him and stuck the nipple of the bottle into his mouth. This was the answer! Here was Cody, confined to his bed and very afraid that his "mother" might leave him. He would scream if I moved so much as a foot away; it was the perfect situation for talking tuition!

I stood up abruptly, pulling the milk from him. Immediately a piping, high-pitched "pi-pip-pirri" burst from his pouting lips. I was down beside him in an instant, giving him both food and praise. I stood up again, and once more came the barrage of anxious chirruping. Success! Twelve times more I repeated the maneuver, and each time, to my increasing disappointment, it was slightly less successful. My hairy pupil adapted to new situations just a little too quickly. Within a few minutes, he was once again as dumb as ever. I even tried hiding behind the playroom cupboard, but each time I peeked out to judge Cody's responses, I found him lying languidly against the side of his bed, confidently awaiting the coming of his food.

There was one last hope. Instead of moving away or hiding, I walked to Cody's left side. Turning his head, and still absolutely silent, Cody followed me with his eyes. I continued leftward, and as soon as I was out of sight Cody began to turn himself in my direction giving anxious little chirrups as he did so. I came back to him immediately, feeding and petting him for all I was worth.

During the next 15 minutes anyone looking through the window of the hut would have thought I was either the victim of an attack of St. Vitus' dance or the inventor of a new slimming exercise. I alternatively jumped over or ran around the bed at an everincreasing pace, trying madly to lose Cody's eyes by the speed of my movements. Cody was equally active, the long mane of red hair on his back swirling out at right angles to his body as he pirouetted on tiptoe, a little spinning top in a desperate effort to keep "mama" in sight. I managed to get nine more squeaks from the breathless brown bundle before I collapsed next to his bed, quite exhausted. It was progress, but in order to continue this particular line of research, I would first have to get myself into training.

I decided for the time being to concentrate on something less strenuous. Today was also the day for analyzing Cody's achievements on the Gesell test, a series of problems devised by Arnold Gesell, the famous American child psychologist. In a way, the Gesell test is like an automobile's MOT test; it discovers any flaws in the "vehicle," testing whether the human infant under observation is sufficiently "roadworthy." There is, however, one important difference. Whereas the MOT test has the same level of excellence for each passing year, with the Gesell system increase in age is expected to bring a definite improvement in the "vehicle's" overall performance, braking distance, maneuverability and handling, etc.

I was giving Cody the motor development section of the Gesell test. This involved determining to what extent he could control himself when placed in a variety of postures or situations. For example, times of attainment of stable sitting, head raising from prone and supine positions, pull to standing using a chair or table, climbing stairs, and many more locomotive patterns are thought useful, and their achievement or otherwise is noted down on mimeographed sheets. Gesell tested more than 400 human infants in his classic study that formed the basis of the test. Each infant was photographed from two angles simultaneously, every foot of movie film being analyzed with great care.

From his researches, Gesell was able to devise a table of "normative" results, giving the "usual times" at which a normal, healthy human infant begins crawling, walking, running, etc. While no child will ever develop all aspects of its behavior at precisely the time given by Gesell, the dates of these developments *overall* (and there are 63 items in the motor section of the Gesell test alone) should tend to cluster around the dates given by this test as normal. If they differ markedly, we can be fairly sure that the infant is in some way unusual. The test is therefore a sensitive screening device. It has given pediatricians the world over a useful tool for detecting abnormal children, allowing them to receive remedial treatment as soon as possible.

But there is another use of the Gesell test, one which it is doubtful Arnold Gesell ever envisaged. By giving the same tests to those animals closest to man on the evolutionary scale—to the other primates, and especially the great apes—it is possible to determine how, and in what way, an ape either excels or falls behind the average human child. Work on the gorilla and chimpanzee has shown that these apes develop much faster than human infants. For example, a human child only learns to roll from its back to its stomach at the age of 24 weeks. Gorillas are accomplished rollers at ten weeks, while chimps take a little longer to achieve this maneuver—at around 11 weeks of age. No one had yet asked the orangutan to compete in this behavioral "motor race," and I was interested to discover just how Cody would fare. I gave my fosterling the motor development side of the test every two weeks, and the results, when I finally analyzed them, were both puzzling and exciting.

While the differences between chimp and gorilla were, on most occasions, kept to a few weeks at most, Cody true to orang form, showed wild variations. He seesawed back and forth. In one case an action would be seen first in Cody, with the chimp and gorilla following later; then, with another action, the positions would be reversed. Overall, however, with many behavioral items, I discovered that Cody developed more slowly than either the chimpanzee or the gorilla, yet still at a faster rate than a human child. "Creeping" showed this well: chimps and gorillas creep at around 20 weeks of age, Cody crept at 28 weeks, while human infants do not employ this particular method of locomotion until week forty.

This discovery, that Cody was the middleman when compared to humans and other apes, excited me. It is generally agreed that a longer developmental span is correlated with greater intelligence, the theory being that the longer the time to maturity, the greater the opportunity the infant (be it human or ape) has to assimilate and learn from new situations, individuals, and objects. So it was possible that Cody's extended motor development implied a potentially high intelligence. If this was so, then Cody's slow progress could be a natural consequence of his more extended development. After all, I told myself, human infants do not really begin to talk until they are twelve months old, and some late developers take much longer. That idea cheered me considerably. I began to feel a little more confident about the future, and especially of Cody's chances of talking.

8

Cody was allowed outside only on very fine days. Most of his time was spent indoors in his thermostatically controlled playroom, his mind engaged in the various tests I imposed upon his waking hours. At around this time a survey of Cody's grasping development was coming to an end, and I could review and compare it with that of human grasping.

We all take grasping for granted, yet it constitutes a very important activity in *Homo sapiens*. Imagine trying to live your life without being able to take hold of things—to lift a cup from the table, hold your knife and fork, plug in the radio, put on your clothes—and you soon realize how indispensable a skill it is. Were visually-directed grasping beyond the wit of man, our ability to act effectively on the environment would be almost completely undermined. Much the same can be said for the rest of the primate order. In all infant apes and monkeys, reflex grasping acts to keep the infant in close physical contact with its mother. Grasping and the manipulation of objects form an integral part of behavior in all adult primates, especially the chimp, gorilla and orangutan. Its main uses are in food collection and locomotion, but other acts, such as using a club for defense in the chimpanzee and branch dropping in the orangutan, are also important.

The act of grasping an object that we see appears to be a relatively simple maneuver: you see a glass of sherry, and you pick it up. Closer analysis, however, reveals that many less complex movements need to be coordinated to allow grasping on sight. Try holding this book away from you in your left hand. Now grasp the top of the book with your right hand. Easy. But to do this you will have had to:

1. Fixate the book with your eyes.
2. Activate your right hand at the sight of the object.
3. Move the right hand from outside your field of vision until both book and hand are simultaneously in view.
4. Direct the right hand toward the top of the book.
5. Open your hand as it approaches the book and close it at the correct moment.

Visually-directed grasping is therefore made up of at least five components. It is a *super ordinate* behavior and, as such, far beyond the coordinating capacity of a newborn infant. This is very hard for an adult to understand. We get so much grasping practice during our daily lives that the action has become second nature to us. But that is just what it is—second, not first, nature. We have to *learn* this skill, and we learn it during a time of our life that few of us remember, the first six months of independent existence.

Many people have investigated grasping in the human child, but it has been perhaps most profitably studied by the Swiss psychologist Jean Piaget. This scientist observed his own children on a day-to-day basis, and with this simple approach produced some profound insights into the early development of children. As far as grasping is concerned, Piaget believes that by a process of interaction with the environment, a quite definite series of grasping abilities, or "levels," unfold as the child develops. To all intents and purposes, this series is thought to be inviolable.

Level 1 is the reflex grasp where an object, say a toy rattle, is placed in the child's palm. Later the child grasps if the object is touched against the finger tips (Level 2). Next, grasping occurs when the rattle and the infant's hand lie close together in the child's field of vision and, following this, when hand and object are distant from one another, but are still simultaneously in view (Levels 3 and 4 respectively). Level 5, the final stage, is achieved when, as in adult grasping, the hand can be brought from outside the visual field and successfully directed to the object.

What is important to realize is that (if Piaget is correct) Level 2 will never be seen before Level 1, Level 3 will never be seen before Level 1 or 2, and so on. If such a series is true for human infants, it may also hold for other primates. If so, we would have an interspecific "law," a rule that holds for more than one species and a great aid for any biological research. It was for this reason that I spent many long hours presenting interesting objects to Cody and noting down his grasping responses.

To my great satisfaction I found that although Cody progressed almost twice as fast as a human child, he was showing an almost identical series of "levels" as he gradually progressed toward full grasping on sight. There were only two discrepancies: hand clasping (a Level 1 behavior according to Piaget) began later with Cody, in Level 2; and as if to compensate for this laggardly development a Level 2 behavior (increased hand movement on seeing a desirable object) began in Level 1. Apart from this, everything about Cody's grasping development obeyed the same rules as those found for human infants. It seems then, that among our closer relatives, the development of behaviors such as grasping are at least partially "wired in," in an identical manner to ourselves. It would be interesting to see if animals lower on the evolutionary scale also show this pattern, but we will have to wait for more research before such a basic ground plan can be shown.

Cody might have accomplished grasping on sight in advance of a human infant, but he was still no closer to vocalizing on command. If anything, he was even further from this ability because, during the months of talking tuition, he had learned to crawl. I had found him out of his bed one morning a good five feet away from it. In order to see how he had managed this I at once picked him up and moved him farther away from his beloved refuge. Until that time, Cody had always merely screamed for help if he wanted to move, holding up his hand in the most imploring and heartbreaking manner. On this occasion, however, he crawled straight for his bed, although from his uncoordinated movements it was obvious that he had only just learned this method of locomotion.

He could not yet manage a straight one-two-three-four movements of limbs, but instead pivoted his bottom over and around his legs, these remaining immobile until his *derrière* had reached *terra firma.* Once he had bottom-landed, his legs would move again to a forward position, and he would once more lift his rear in a little over his legs. His later efforts were also a bit of a hit and miss affair in that he would attempt to raise all four limbs together (no one had told him about gravity) and, with a little cry of indignant surprise, fall flat on his face. Within two weeks he had discovered the most efficacious form of locomotion on four limbs, and from then on my life was hell. I couldn't move an inch without Cody scurrying after me, no thought in his tiny mind other than getting hold of me and not letting go.

His crawling was not without its entertaining moments, though. One day, while following me from playroom to workroom, Cody, like the genius he was, took the shortest possible route—beneath a chair on which several pieces of his wardrobe were hanging. As he passed out from under this, a blue "Babygrow" suit fell from the seat and landed directly on top of his head. He began crying at once and spent the next few minutes walking round in a small circle, one hand on the floor, and the other pulling vainly at the object that had "turned off the lights."

But the worst aspect of Cody's new-found skill was that, instead of producing a sound when I moved away from the bed during his talking sessions, Cody simply climbed out of it and followed me. He often screamed as he approached, and I was quick to reward him. But his terror at being more than three feet from his adopted parent seemed to drive any "rational" thought from his mind. He really was scared until he reached me, and so the connection "scream equals food" simply was not made. I had been trying this new idea for about three weeks now, and we had not made an inch of progress. I could not wait forever for a talking breakthrough. My research grant would not last indefinitely: I would have to try to find a solution on my own. And quickly!

In desperation, I tried a new tack. To speak, bark, whistle, or do anything vocal, it is necessary to expel air from the lungs by way of the mouth rather than the nose. So, as Dr. Furness had done with his orangutan sixty years before, I decided to teach Cody to breathe through his mouth. But this skill was something Cody seemed unable or unwilling to learn, for whenever I opened his mouth with my fingers he continued to breathe through his nose. I tried sitting him on my knee facing me and putting my fingers across his nostrils, blocking off this particular method of breathing and forcing him to use his mouth. But Cody was so scared by this cutting off his air supply that he would try to squirm and wrestle my fingers from his nostrils, at the same time clamping his mouth tight shut. I could not hold it open—Cody had extremely strong jaw muscles and, by now, a growing number of very sharp milk teeth. Unable to breathe either by nose or mouth, the little ape would gradually begin to asphyxiate. So afraid was I of doing him serious harm I simply had to release my grip on his nose, especially when his normally pale face began to turn bright scarlet and he still refused to mouth-breathe. I was quite at a loss: should I keep on with this training, return to the scream-and-feed routine, or try something completely new? But what? I was still in two minds when disaster struck.

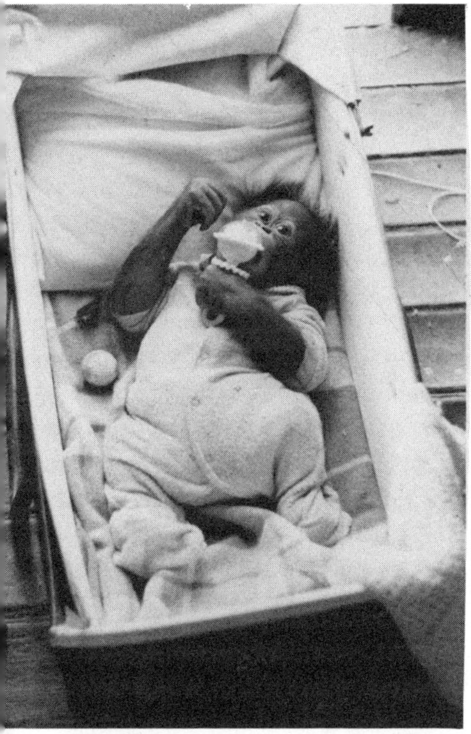

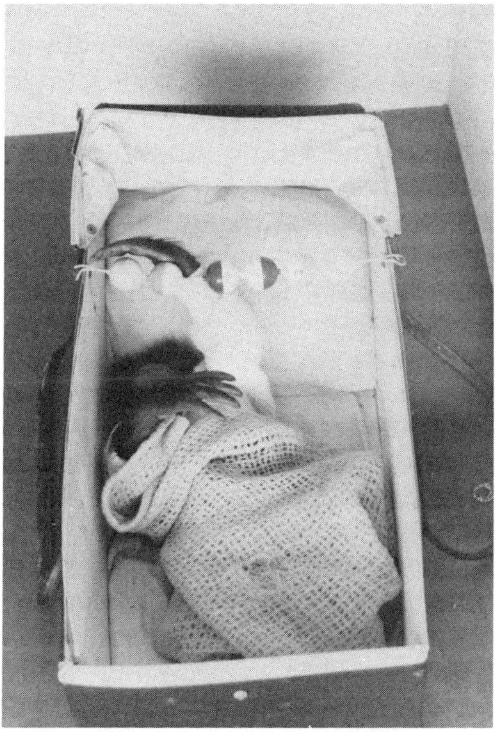

I wanted Cody to have all the pleasures of a human childhood—including toys and tickling.

Cody liked to let me build with his play cubes so that he could immediately knock them down.

Cody was at first fascinated by his own reflection, but he soon got used to the mirror. *(John D. Drysdale, Camera Press Ltd.)*

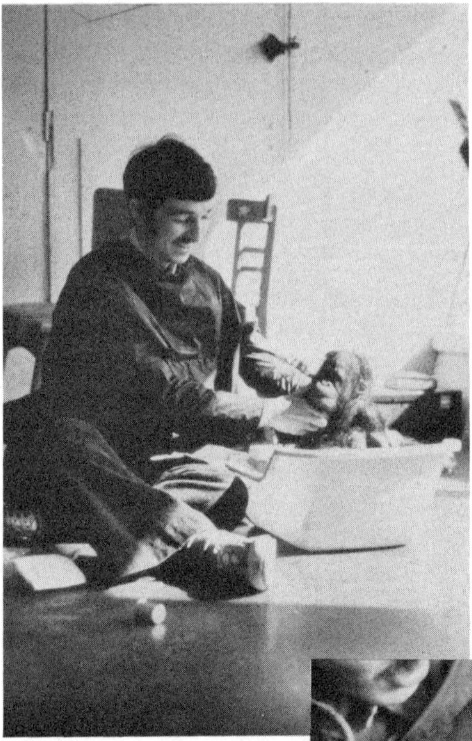

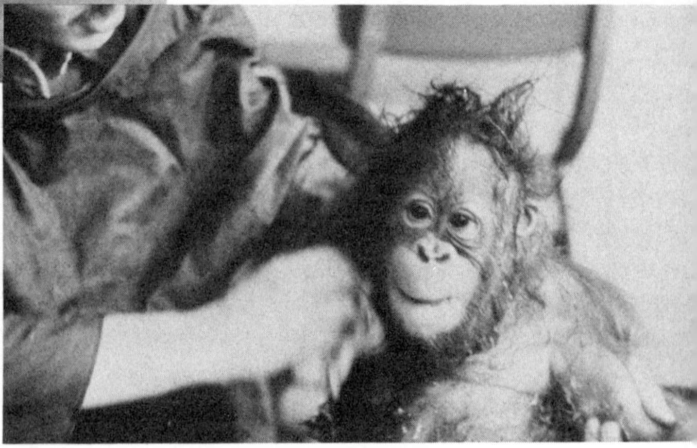

Bathtime was a twice weekly affair which Cody hated unless he managed a bite of his favorite treat—the soap.

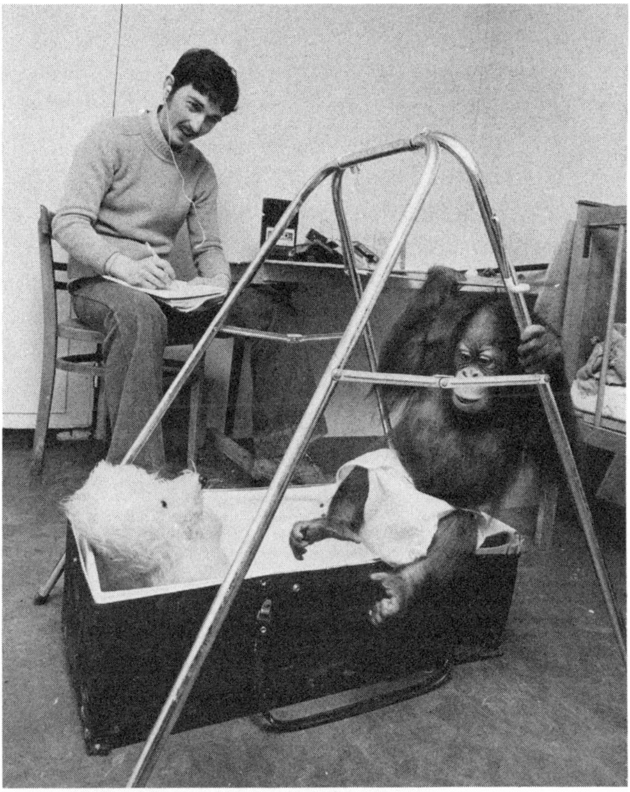

Cody exercising at 9 weeks. *(John D. Drysdale, Camera Press Ltd.)*

Instead of crawling, Cody often used his climbing frame as an aid to standing and walking on two legs.

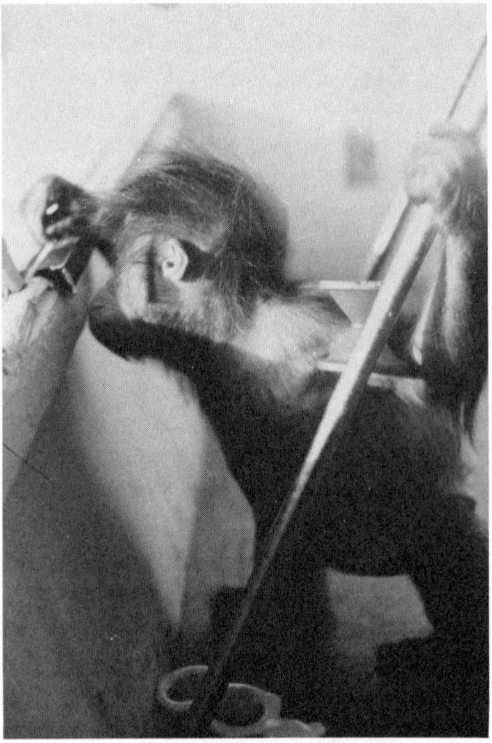

With the help of the climbing frame, Cody was able to explore much further afield—without the risk of collapsing in a heap on the floor.

Playtime was a very boisterous affair and very hard on Cody's toys.

Cody was growing up into a very handsome Orang.

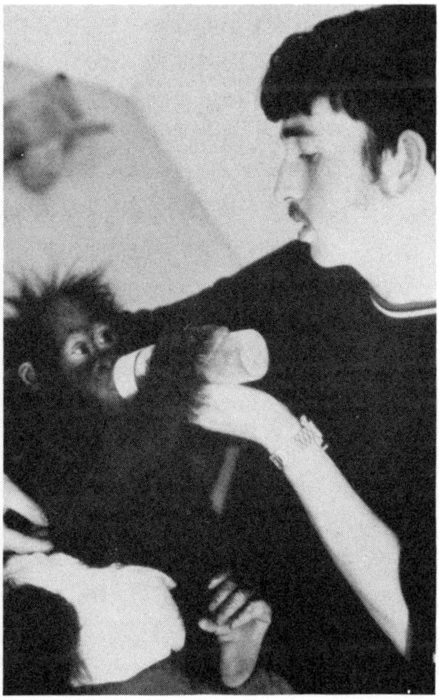

Cody loved milk—he simply took hold of the bottle and never left off untill the last dreg had been consumed.

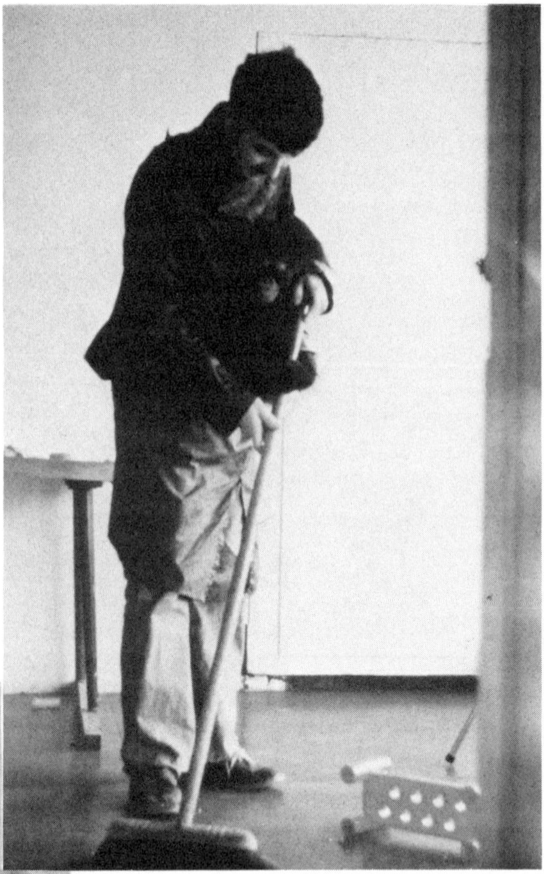

Cleaning up became doubly tiring when Cody demanded to be carried everywhere about the hut.

Lots of people came to watch Cody growing up, though I often wondered just who was observing whom.

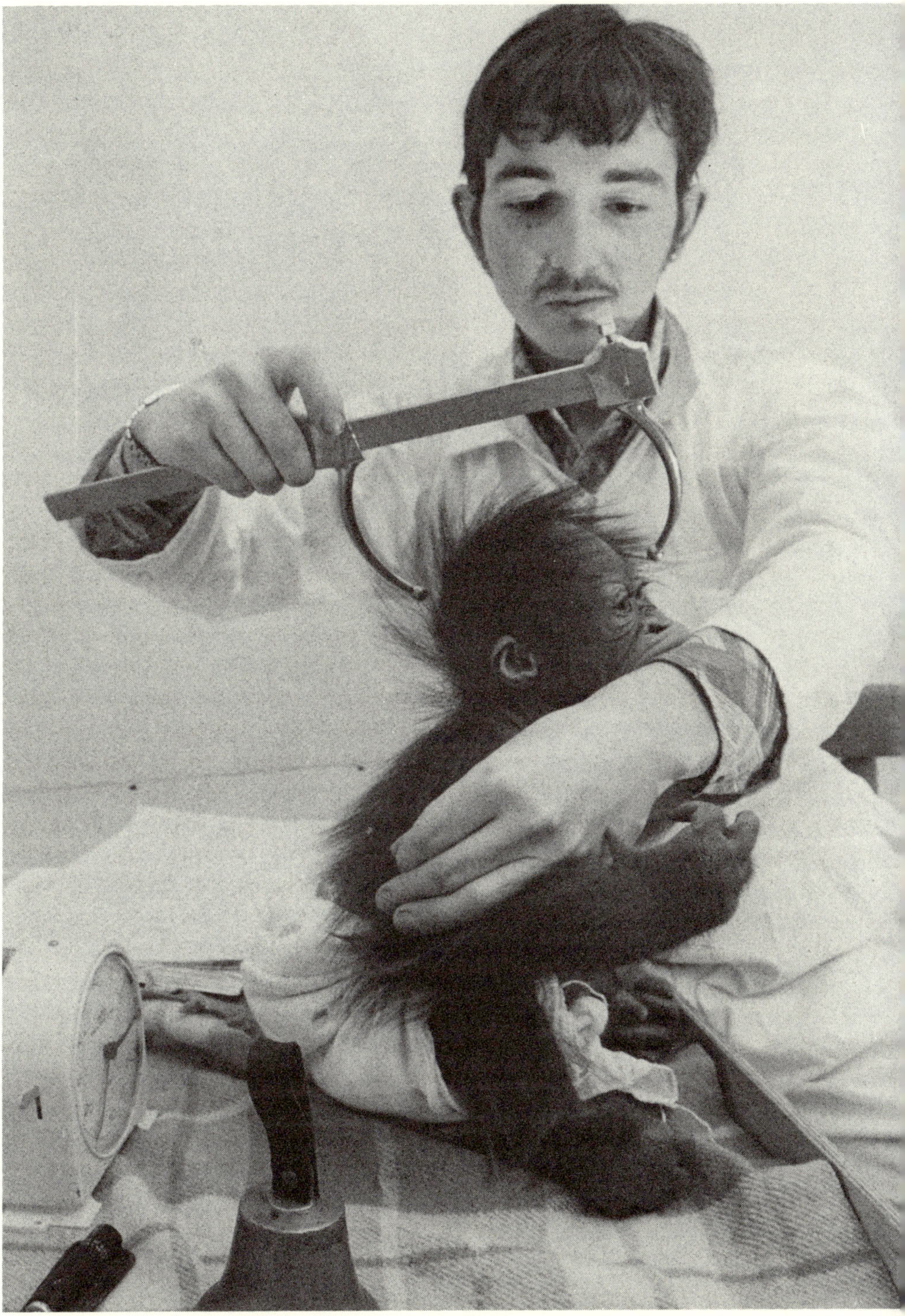

Measuring Cody's physical development became more difficult as he grew older and stronger. *(Ian Tyas, Keystone Press Agency)*

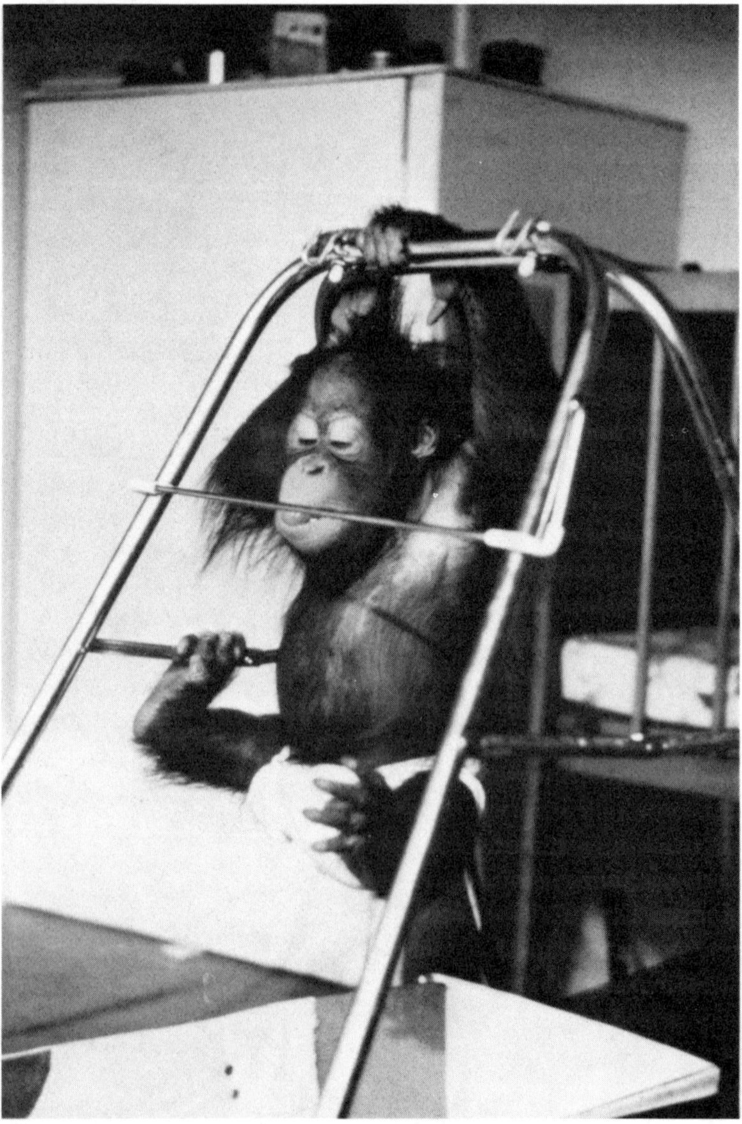

The climbing frame gave Cody ample scope to develop his gymnastic talents.

9

Cody became ill very rapidly; I left him in perfect health and returned 36 hours later to find him in a desperate condition. As soon as I entered his hut, I was overwhelmed by a ghastly stench. In less than two days Cody had contracted gastroenteritis—severe diarrhea—a very serious disease in apes, and especially dangerous during infancy. Though it would have helped the smell enormously, I decided against opening any windows—it was much too cold. And anyway, there was no sense in trying to take gasps of clean air every few minutes, it only made the smell more pronounced. It was probably better to breathe deeply until my nose adapted to the stink, which in those first moments was so strong I could almost taste it.

Cody was nowhere to be seen. Normally, ever since he had been able to crawl, he had always shown his excitement at my return after a few days' absence by coming to the door and welcoming me with a "greeting piddle"—a golden stream of urine. But today I found him lying in his bed, clinging faintly to his security blanket, weak, morose, and altogether exhausted: a sadly different orang from the one I had left just three days ago. His diaper was useless in these circumstances, and it had been removed, probably by Nina, who cared for Cody during my rest days. He had had a bowel movement quite recently, a yellow semi-liquid that his squirming had smeared across the bed's mattress and all over his lower body. He was lying facing the wall, and not until I got very close to him did he turn and look up.

It broke my heart to see his little eyes light up and a feeble smile come to his lips. He chirruped weakly and grasping hold of the bed's top rail, slowly and painfully pulled himself upright. Standing now, his body dripping those evil-smelling drops, he held out his hand in a mute appeal to be picked up and cuddled. The smell and the filth were appalling, but there was nothing I could do but comply. As soon as he was on me, he wrapped his arms around my neck, his legs around my waist, then pulled his face close to mine and seemed to try very hard to give me his best smile.

It was time for Cody's evening meal—a dish of cereal followed by milk. The curator had told me that if he did succumb to the runs, I should add to his cereal at least a teaspoonful of Kaolin powder to dry him up, and a capsule of the antibiotic Penbritin, to fight the bacteria producing the diarrhea. I should mask their rather bitter taste with plenty of milk and honey.

Cody would not touch it. As usual he took one tiny taste, disliked it, and from then on refused to have anything at all to do with it. His head twisted at the approach of the next spoonful, and he fastened his mouth so tightly shut that I could not even prise it open with the metal handle of the spoon. After ten minutes I was forced to give up. But I didn't dare. Until and unless he was treated for this illness, he stood a very good chance of becoming progressively weaker over a period of days and finally dying.

I tried him next with milk. Having lost so much liquid, Cody was a little keener for this and consumed the first fluid ounce greedily. Milk seemed to be the answer. I hurriedly poured more Kaolin, Penbritin, and honey into the top of the bottle, screwed down the teat and tried again. To my intense relief, the small, smelly baby on my lap once again took to the bottle voraciously. It was only when he refused the last two fluid ounces that I realized the mistake I had made. Kaolin powder does not dissolve in milk, and it had formed a thick clay-like layer at the base of the bottle. No amount of pleading or cajoling could persuade him to swallow so much as a mouthful of this gray-colored silt. However, he had got the antibiotic inside him now, and that, I felt, must surely do him some good.

I cleaned up the soiled bed as best as I could with warm soapy water, then turned my attention to the tiny bundle of misery hanging from my neck. A quick bath and Cody began to look a little better. But he still seemed quite dulled to the world; and when I laid him gently on a clean blanket in his bed and made for the door, he did not even cry after me.

It was then that I became seriously worried. He really was ill. Extremely apprehensive as to what the next day might bring, I closed the hut door quietly behind me and trod a doleful path down to my trailer.

It was with a great deal of trepidation that I approached Cody's room next morning. I had made the long walk up at nine and again at twelve o'clock the night before and, although the odor in his cage had not been that of roses, the patient had seemed to be sleeping quietly, and I thought it best to let him rest.

I opened the door and once again the fetid stench of diarrhea hit my nostrils. Walking to the inner door I felt like giving up there and then. Cody was still in his bed, on his blanket, just as I had left him. He looked up at me, a small spark of pleasure just visible in his rheumy eyes, but no movement of limbs or body followed. He was gradually becoming weaker, and if something was not done soon he would certainly die. Perhaps within the week. I had sent for the vet, and he was supposed to be coming some time this morning. But I knew that the important thing was to maintain Cody's will to live. He needed every bit of love, affection, and kindness I could give him.

Cody came to me readily. I picked him up and put him on my right hip, his favorite position, but his little hands were so weak I had to help him hang on to me. I had not had to do *that* since he was two-and-a-half months old. The bed was once again smeared with the horrible ghastly-smelling mess, I bent down to clean it up and at the same time heard three splatterings. On the floor were three mounds of what looked exactly like cake batter—*before* you put it in the oven.

I left the bed and turned my attention to the floor, carrying Cody through to the workroom for the bucket and mop. As I did so, I heard again the ominous plitter-platter. Spinning round I could see a trail of very fluid yellow feces extending from playroom to workroom, and worse still, he had done it again as I turned, the centrifugal force flinging out the feces all over the walls and windows! I set to work immediately, but it was like a nightmare—no sooner was one spot cleaned than Cody would perform in another. The stench was ineffable.

My mind was in a turmoil. I hurriedly decided to give Cody his breakfast first, with plenty of Kaolin to dry him up. But on last night's evidence he would again refuse it. Trickery was the only answer. I quickly prepared two pans—Cody making free once more during the preparations—one pan containing ordinary, extra sugary porridge, the other porridge and Kaolin. I offered Cody a teaspoon of the former. He gave it the usual trial by taking in a very small portion and, charmed by its sweetness, swallowed down the whole lot. I gave him another ordinary spoonful, then slipped in a Kaolin-packed portion. He'd swallowed it down before he realized what it was.

Ill he might be, but Cody still knew he had been betrayed. Resentful eyes glowered up at me and the next ladleful—of ordinary cereal—was treated like the first—very, very cautiously. However, another good spoonful, and he was once again eating with gusto, so I slipped a second Kaolin teaspoon in. Yet again it worked faultlessly, and although I had gradually to increase the number of "good" to "bad" portions, eventually Cody had unwittingly taken his full quota of Kaolin. The antibiotic went in with the milk, and this was consumed with no trouble. I felt a bit better. At least I was doing something constructive, and the vet should soon be here.

It was not until three o'clock, after Cody's second feed, that the vet arrived, with John Laing at his heels. Willy, a bright young

Scotsman with gold-rimmed spectacles and a gold-capped tooth, stepped through the door carrying a tube of antibiotic which he said would soon destroy the bacteria that was undermining Cody's health and verve. He was quite confident that the little ape would soon recover, and his optimism helped to lift my own low spirits.

I was given a large white plastic syringe filled with antibiotic, and instructed to squirt it down the infant's throat four times each day. How this almost impossible feat was to be accomplished was left up to me, and after taking Cody's temperature, which, frighteningly proved to be above 101°F, John Laing and Willy swiftly left the smelly scene. For some reason I still felt that Cody was a lot iller than the vet suspected. I decided to bring my camp bed into the hut and to spend the next few nights as close as possible to my patient.

Three nights later I was still in Cody's hut. Willy's prognosis had proved to be overoptimistic, and Cody was now in a desperate condition. In all that time the little invalid had taken no solid food and only a few fluid ounces of milk. Kaolin and morphine had no effect, nor did the oral doses of antibiotic that I administered four times daily. He was losing weight steadily, over a pound in the past two days, and if this loss continued there was little hope that he would live beyond the next week. Even the curator had begun to fear for his life. We had tried everything, and Cody was as sick as ever.

Lying in the camp bed a few feet from the infant's bed, all I could hear was his slow, rather labored breathing and, every so often, a slight twitch as the cramps swept yet again through his emaciated body. Earlier that day, as a last resort, the vet had taken a sample of Cody's watery feces and sent it to a nearby laboratory for analysis. Late in the evening he returned with an antibiotic which he claimed was specific for the type of bacteria attacking Cody. He very gently lifted the infant from the bed, sat him on his lap and carefully inserted the syringe under the skin at the back of Cody's neck. It was a large dose; his neck swelled visibly as the antibiotic went in. To any human the injection would have been very painful, but Cody didn't even flinch. It was a bad sign—the little orang was past caring.

For the rest of the day and a half of the night, I had watched the young ape and worried, waiting to see if his decline would continue, or for an indication that the antibiotic was working. But in all those hours there was no sign one way or the other. I knew that we were doing all we could for Cody, but still the feeling of utter helplessness kept me wide awake. It was 3:30 in the morning when I got up from my camp bed and stood frowning anxiously out of the window.

Then suddenly, I heard a little gurgling cry. At first I thought Cody was about to give up the ghost, but on a swinging around I could see that he too was wide awake! He was standing upright, hanging on with both hands to the top rail of the bed. He was very shaky, but as I approached he lifted one arm and chirruped weakly, his way of asking to be picked up and cuddled. Once he was on me I could feel that his muscle tone was much better, he seemed more tense, much more alive. His eyes, too, had lost some of the dull gaze I had become accustomed to over the past few days.

Would he take any food? The question was of vital importance. Without liquid to counteract the dehydrating effects of his diarrhea, and without nutrients to maintain his bodily functions, Cody might still die. In feverish haste I made a bottle of baby milk and fearfully applied it to his lips. To my intense relief Cody began to suck. He drank only in fits and starts and with a terrible slowness—but by the time he stopped the level of milk had fallen by almost half. Cody looked around dreamily, gave a little burp to show his appreciation, and fell fast asleep!

I put the little invalid back in his bed and tucked in his blanket. I was overjoyed, the drug was working. Willy would be back in the morning with another shot of antibiotic and, with plenty of care and attention, I was sure the young orang would pull through. Cody was on the mend.

10

When, two weeks later, Cody was again fit and well, I had to take stock of his speech-training program. His illness had brought all talking research to a standstill and prospects for a successful restart seemed very gloomy indeed. I did not realize it then, but the whole dark episode was probably a blessing in disguise; it forced me to look long and hard at the talking techniques I had been using.

Before his illness I had got nowhere with the two methods of tuition that I had been trying out on Cody. There was little to be gained by starting where we had left off. In more than three months of teaching the young ape did not seem to have moved one syllable toward either voluntary vocalization or mouth breathing, the first two subgoals of speech. This was not to say Cody had learned nothing—the thought processes of apes are just as covert as those of man, and it is more difficult for us to "read" ape expressions than those of humans. It was possible that Cody had processed all the information I had been feeding him and that he was on the verge of speaking. If I began the scream-and-feed routine, it was conceivable he might be vocalizing within days. But I simply could not be sure, and it was equally possible that this particular training could be continued indefinitely without any breakthrough at all. It was, I felt, time for a change.

At the risk of being thought mad, I decided to take my problems to the experts, to the speech therapy department of Newcastle University. It was a great relief when the principal, Mrs. Margaret Edwards, expressed no surprise at my request to visit and discuss Cody's "problem." Indeed, she was exceedingly interested in the possibility of "ape-talk," and we arranged an appointment there and then for the following week. At this meeting, I was given many valuable suggestions, and it was one of these that finally proved to be the key to successful vocalization.

At the end of a two-hour talk, which centered primarily on Cody's personality and temperament, Margaret Edwards suggested that the best method was probably to try one of several regimes designed to teach speech to autistic children. At the time I knew only that autism was a form of mental dysfunction that afflicted children. How, I wondered, could the experience gained with these children help Cody?

To the Newcastle experts it was all very clear: it seemed that my description of Cody's personality resembled in many respects that of an autistic child. Techniques that had helped such children could, they reasoned, aid Cody. It seemed feasible, but before I accepted their advice I had to be sure their assessment was correct. I left Newcastle with new hope, many references, and a determination to find out as much as I could about childhood autism.

An autistic child may appear quite normal, as I discovered on my first visit to an autistic center. The children are often attractive, they move well, have good coordination and, although perhaps a little pensive, they seem intelligent enough. But this superficial normality masks a very special form of mental retardation. Autistic children are unable to relate to other people. Even after infancy they seem incapable of forming attachments, even with their parents or siblings. Mutism (the complete absence of language) is found in approximately 50% of all autistic children, and those who can speak often have language difficulties. Echolalia (echoing the speech of others) is common, as is the confusing trait of pronoun reversal, where the child might say "he's here" or "he knows that" when what he really means to say is "I'm here" or "I know that."

As I mentally compared Cody with these children, I could see that there was much in Cody's behavior that did resemble an autistic child. Happily, the young orang did not display the grosser symptoms of autism. Autistic children often mutilate themselves—they have been known to bite themselves until they bleed—and they often perform stereotyped movements such as unceasingly rocking to and fro. They may spend hour after hour manipulating mechanical objects and throw temper tantrums if their daily routine is slightly changed.

But like most of these children, Cody was mute. He was also quite introverted: self- rather than other-directed. For example, Cody would very often play with me, but at other times he would ignore me completely if I tried to initiate a game or called him over for a tickle. On these occasions it seemed he preferred to play by himself. I had taken this to be a sign of Cody's inborn "orangness" (because of their mother's solitary habits baby orangs in the wild rarely have the chance of playing with infants of a similar age) but it was still very similar to the closed off attitude of human autistics. Likewise, Cody's attitude to me was, at times, very restrained. He *was* devoted to me it was true, but more as an object of comfort than as a person. When the mood took him, he would move heaven and earth to hang himself around my neck, but having achieved this coveted position, he was more often than not completely indifferent to my smiles or expressions of endearment. His behavior was probably fairly typical of any baby orang, but it was also close enough to an autistic child's for me to feel that autistic training methods might prove effective.

I began reading the speech therapy articles Margaret Edwards had given me, and the more I read, the more convinced I became that here was a method that might mean the difference between Cody talking or remaining mute. All the methods were rooted firmly in basic operant conditioning techniques devised by B. F. Skinner and his disciples. Such methods are intended to modify behavior toward a desired goal by means of positive or negative reinforcement. (For positive reinforcement read "stick.")

To Skinner and his colleagues all behavior, including human behavior, is controlled by our expectations of reward and punishment. The reward does not have to be concrete; an ape (or a child) might work for candy, but a martyr will cheerfully burn at the stake for his own special reward—that of knowing he is doing the right thing. This, at least, is the credo of the Skinnerians. Few other scientists are totally committed to this "stimulus-response" view of life, especially human life. But it cannot be denied that operant conditioning with less sophisticated organisms such as pigeons or human infants had produced some remarkable results. By rewarding pigeons each time they made a move to the right, Skinner eventually had the poor birds walking in circles or figures of eight. He even had a scheme for kamikaze pigeons. Imprisoned in the nose of a flying bomb, the (suitably trained) bird was to direct the bomb on to its target (and itself into oblivion) by pecking at directional screens!

In many ways, Catherine and Keith Hayes's experiment with their chimpanzee Viki was pure operant conditioning. But it appears they were mistaken in taking the most obvious route. They wanted Viki to speak, so they rewarded her for any vocalization in the hope that they could gradually shape the sound into an acceptable word. In contrast, the new "autistic" methods depended on imitation for their efficacy. More, they began with *physical* imitation and only when this was well established did they pass on to vocal imitation. Such "aping" methods seemed ideally suited to a young orang and, if available at the time of Viki's experiment, would probably have produced better results in the chimpanzee.

One account specifically mentioned Viki, pointing out that when autistic children were taught with a "shaping" routine (as Viki the chimp had been taught), they showed only as much progress toward talking as had the chimpanzee. Yet these same children, when given an operant conditioning "imitation" routine, had learned up to 250 words, an achievement that had allowed them to break out of their silent mental prison and become at least partially functioning members of their families.

I was eager to find out if Cody would respond as well as an autistic child to imitation training. But first I had to pick one method from the score of variations on the basic imitation technique. To be so close to the solution and to choose wrongly would be disastrous. After much thought I decided on a variation used to good effect by Dr. Frank M. Hewett. The Hewett method had two advantages: not only was it known to be successful, but also it could be set up with the limited materials and resources available to me. Other methods demanded very sophisticated and expensive equipment, everything from headphones through amplifiers to cathode ray oscilloscopes. The only apparatus Dr. Hewett had used was a large wooden cubicle.

Hewitt had taught Peter in a two-sectioned "talking booth," teacher and pupil communicating through a cavity which the teacher could close by a shutter. Only the teacher's compartment was lit. For each correct response Peter was rewarded through the shutter with food, drink, or music. Punishment (when he refused to respond, began screaming, etc.) was the removal of light and personal contact, the shutter being closed for this purpose.

Peter worked well under these rather Draconian conditions, soon learning to imitate such actions as "hand on head." Next he was required to imitate sounds; and the final phase was to teach Peter to use his sounds in the "external environment" of everyday life, an ability which he mastered with surprising speed. Within a few weeks, and for the first time in his life, the boy was using words and phrases in conversation with his family.

At first sight it may appear cruel to lock a young child in a small cupboard without light or food. But as Peter's subsequent behavior showed, this really was a case of "being cruel to be kind." Peter learned extremely rapidly: he mastered some 150 words during a

14-month period of intensive training. And the techniques seemed to initiate Peter into an awareness that language existed, and that it could be used to communicate. According to Dr. Hewett, after his period of intensive tuition Peter showed "an insatiable desire to learn new words and phrases."

Transforming a silent child into a speaking one verged on the miraculous, and this unusual technique was clearly immensely effective. But there was still no guarantee that it would work when I tried it out on Cody. For one thing, humans normally do speak, orangs never do. I remembered how Beth the dolphin trainer had vouchsafed her "infallible" technique for producing singing dolphins—Cody had short-circuited *that* little plan. I could not help wondering what perverse and ulcer-provoking plans the little ape might devise when I asked him to put his hand on his head or to imitate placing his finger on his nose. Could he ever be as accommodating as Peter?

11

In the wild, the orangutan has a wide variety of natural vocalizations. John MacKinnon, who spent many months studying these apes in the forests of Borneo and Sumatra, recorded upward of 15 separate calls including kiss squeaks, grumphs, gorkums, lorks, raspberries, shoors, barks, chomps, play grunts, and various screams. Most of these inscrutably named calls serve purposes which, at present, we can only guess at; however, MacKinnon did discover the probable meanings of several. Some of these are perfectly understandable to humans: an orang who is in pain screams or moans in much the same way as *Homo sapiens*. Other calls mean the exact opposite of similar sounds used by man. An orangutan who brings his pout-shaped mouth to his wrist and proceeds to give a long lingering kisslike sound in your direction is *not* making amorous advances but telling you to go away and leave him alone.

Cody, as he matured, would probably develop at least some of these natural calls, and this posed a not inconsiderable problem. Just supposing I did teach the young ape to speak, and that some of the sounds resembled the orang's wild calls; would not the sceptics say (and with some justification), that Cody was simply "doing what came naturally"? It was a problem I had to avoid. The sounds Cody was taught would have to be quite different from the "ape talk" of his wild brothers and sisters.

I decided from the outset on four principles that would dictate the functional approach I wanted to take toward Cody's "ape speak."

(1) Each sound Cody was taught should designate a specific referent (i.e. object) or class of referent. Thus, he would not be allowed to use word A and word B for the same object. Each had to have a specific meaning.
(2) The sound Cody used for any object should, wherever possible, follow the first phoneme of its English equivalent, e.g. "fuh" for "food."
(3) The sounds taught would be arbitrary and chosen for their ease of articulation and tuition. This was to produce a bias toward unvoiced sounds.
(4) I was not expecting Cody to produce fully articulated words—only distinct, consistent, and understandable sounds. As I was not sure Cody would speak at all, it seemed rather pointless to teach say, a "fuh" and a "duh" sound then to combine them to give "food," when a simpler "fuh" would serve equally well.

Cody's reaction to the new-style training was very encouraging. As ever at the start of any experiment there were teething troubles with the apparatus. Cody once again demonstrated the difference between himself and human infants, for he did not take at all well to the double-cubicle talking booth. He remonstrated long and loud at the imposed separation from "mother," venting his anger by gnawing at the dividing wood panel with such vehemence that the booth shook and plywood splinters sprang into the air.

I had hoped that the promise of "buying" contact comfort with a correct response would have helped the training program. But Cody was so uncontrolled and hysterical in his attempt to reach me it became obvious that he would make none of the mental associations I wished for while placed in what for him was a very traumatic situation. So down came the dividing wall and we started again, this time with Cody firmly attached to his surrogate mother.

Eventually we settled upon a talking booth of more modest design. It was essentially a large plywood box some six feet tall by five feet square. I sat cross-legged in this outsize tea chest, with Cody facing me on my lap. Although rather crude conceptually, this "experimental apparatus" proved a very satisfactory training environment for the infant. It cut off all outside stimulation, centering Cody's attention on my hands and face, and it also allowed me to manipulate his hands and mouth into the position required for imitatory movements. Best of all, it stopped Cody screaming.

I may have modified the booth for ape teaching, but the reward for a correct response remained utterly human. For each approved action Cody received a Cadbury's Chocolate Button! And a very satisfactory reward it was too. Cody loved chocolate with an undying passion, and as he was rarely given any of the smooth brown confection, it was the one thing he would work wholeheartedly to attain.

Phase One of tuition—the introduction—was simply accustoming the little ape to the pleasures of the talking booth. I spent the first two weeks of training sitting with Cody in the booth for two half-hour periods each day, playing with him and feeding him a chocolate button every five minutes or so. Cody must have thought he was in paradise.

Chocolate proved something of a superweapon in the fight to train Cody. Within seven weeks of ending the introduction phase, Cody had learned his first imitatory movement—"hand-on-head." Initially there had been no indication that he understood what was required of him, only that he greatly resented the loss of his play-and-chocolate period. In this second phase, I first demonstrated the movement to be imitated, saying "Do this!" placing my hand on my own head and then leading Cody's hand through a similar movement. As soon as Cody's hand touched the top of his head I explained "Good boy!" released his hand, and quickly popped a chocolate button in his mouth. Cuddling and tickling—which Cody enjoyed almost as much as chocolate—followed, and we were then ready to begin the imitation sequence once more.

This was the ideal sequence, but what actually happened was a little more complicated. Despite the chocolate, Cody fought hard to prevent this invasion of his freedom. He did it by simply "seizing up"—locking his arms close to his chest, tensing his whole body, and refusing to budge. At the time Cody was almost eight months old—he stood just over one foot tall (16.8 inches from head to rump)—yet even at this tender age and diminutive size, he was almost as strong as I was. It took all my strength to move his reluctant arms from their position on his stomach to the top of his head, and I make no apology for the fact that my "Good boys!" during this period were a little short on sincerity. Passive resistance was bad enough, but during the first three weeks of phase two Cody would often become quite aggressive. Sometimes he became so agitated by the enforced armwaving that he would take vengeance on me by biting or striking out. On one occasion his vegeance was not only painful, but extremely embarrassing.

The talking box was clearly visible to outsiders through the large glass window of Cody's cage, and it was this simple fact which led to my discomfiture. On the day in question Cody was being very obstinate about training. He wasn't hungry and, on top of everything, he had suddenly found a new source of interest. My nose! Now I have never thought of this particular part of my physiognomy as a crowning glory. I have occasionally felt quite self-conscious about its more-than-modest proportions; but it had never before endangered my life.

As I patiently modeled the hand-on-head action for the umpteenth time Cody swung his hand in a beautiful right hook, catching me on the bridge of the nose. The world exploded. I crashed sideways, breaking open the booth door and collapsing on the ground, blood spurting from my nose.

As I got up, I saw through watering eyes that the whole episode had been witnessed by a small aged matron, resplendent in a black beaver lamb coat and that mandatory accessory of all old ladies, a black felt hat. She was staring in shocked disbelief at the bloody appartition that had suddenly burst into what had been an empty cage. Keeping a tight hold on Cody and an even tighter grip on my nose, I staggered to the door and opened it intending to explain.

"I'm drying do deach him do dalk," I offered nasally, "but he's hid me."

She tried not to show it, but the old dear was genuinely frightened, her bright eyes darting this way and that as she searched for some means of escape. The young man with the monkey was obviously mad. He might even be violent.

"That's nice, love," she said, "very nice." She looked about agitatedly. "Well . . . I must hurry if I'm to catch the bus."

I never knew old ladies could run so fast.

When, having stemmed the flow of blood, I returned to tutoring my wayward infant, he was just as uncooperative as before. Obviously a bit of wholesome discipline was required. I tried wrist-slapping, but Cody continued to use his arms like the sails of a windmill. A smack on the bottom was treated as an invitation to wrestle play, and even a sharp crack on the head did nothing to cow the diminutive destroyer on my lap.

By this time, I was at the end of my tether; it seemed as if I was losing control of the training situation. How could I maintain discipline? Thinking back, I recalled that with autistic children, refusal to make eye contact with the child is often enough to prevent antisocial behavior. I tried it with Cody looking steadfastly to my left while I sat, like a block of stone, hands and legs crossed to preclude even a hint of friendship.

Cody was completely indifferent. My withdrawal from play meant nothing to him, and I realized I had not taken into account the orangutan's inborn predisposition for autoplay. Mother is the orang child's only friend, and she is not always in the mood for fun and games. So the infant orang, unlike the supersocial chimpanzee, is designed *not* to crave social play. If another infant happens to appear the two will frolic together as they hang 200 feet up above the forest floor. But otherwise the baby orang is quite content to find his own amusement.

It was the same with Cody during my "withdrawal" periods. As long as I was within grasping distance, it did not matter too much to Cody if I did turn to granite. I rather suspect he enjoyed my self-imposed immobility, for often, as soon as I averted my eyes, the red terror would proceed to use me as a gym, swarming up my pullover using my ears as handgrips until he sat triumphantly on top of my head. Then he would slide noisily down my back, only to recommence the ascent the moment he hit the floor

I could not blame Cody for these antics—after all, infant orangs had been enjoying themselves climbing and swinging for more than a million years. Cody wanted nothing else. But *I* did, I wanted him to speak: and I reasoned, perhaps Cody would too, if only I could show him how useful it was.

After several such wild sessions I decided it was time to bring up the big guns—to use Time Out from Positive Reinforcement. With this procedure, the miscreant is removed from any sort of pleasurable experience (positive reinforcement) for a set period of time (time out). Time out increases on each occasion a misdemeanor is committed. If, for example, light is the positive reinforcement, the first time the child becomes "awkward" the light will be switched off for 15 seconds; the next time for 30 seconds, then 45 seconds; in other words, time out is cumulative.

With Cody, movement and play seemed to be the most pleasurable experiences during talking tuition; so these would be denied him. The first time Cody acted up, I pinioned him to the floor, holding him there for 15 seconds. The next occasion would merit a 30-second stay. To term this rather horrible treatment "Time Out from Positive Reinforcement" seemed to typify the unemotional euphemistic approach of most psychologists. I called it a Punishment Period. I did not like treating Cody in this way, but luckily I did not have to do it for long.

When first subjected to a punishment period, Cody screamed with fright and anger. It needed all the force I could muster to hold him down, but the experience had the desired effect. He was so averse to this form of discipline that from then on I had only to threaten the little one with a "P.P." and he would sit with rapt attention while I intoned my never-ending exhortations to "Do this!"

Once Cody had accepted the fact that he would have to live with my forcing his hand on to his head, progress began to speed up considerably. It took four weeks of "Do this!" "Good boy!" and Chocolate Buttons before I felt a very slight decrease in Cody's resistance at the end of the hand-on-head movement. I persevered; and one week later, after 200 hours of tuition and 53 packets of Chocolate Buttons, Cody had mastered his first imitatory action. All that was required was for me to say "Do this!" place my hand on top of my head, and Cody would follow suit. Now, at last, I knew Cody could learn to imitate *movements*, an important stage on the way toward *sound* imitation.

But an even more interesting fact emerged spontaneously, two days later. I had prepared Cody's bottle and was carrying him to the talking booth when, with no prompting whatsoever, the little orang whipped up his hand and placed it neatly on his head. This was quite an achievement: without tuition to "implant" the idea in his mind, Cody had begun to use his hand-on-head movement as a *sign*, a sign that amounted to a request for food. Cody was not simply imitating, he was, in a very basic way, conversing with me: he was asking me for his milk! And of course he got it, with the result that as soon as I removed the nipple from his lips, up would flash Cody's hand to plonk resolutely down on his head in "payment" for another sip of milk. The implications latent in Cody's comical behavior were enormous. To my knowledge, this was the first time an orangutan had been seen to use a hand sign. Chimpanzees had used signs before, both in the wild and in a home-raised situation, but not the orang. And Cody seemed to have grasped the idea of signing by himself.

But Cody's new-found signing skill placed me in something of a quandary. To allow him to request his milk in this way would mean changing the direction of my research from vocal to sign language. The latter was enormously interesting, and I have no doubt that having initiated signing himself, Cody would have taken quickly to the manual mode of communication. But I had set myself the task of teaching the orang baby to speak. Signing and speech were uncomfortable bedfellows; I was sure that being simpler, the former would dominate Cody's behavior completely. There was no alternative—signing would have to go.

I took to ignoring Cody whenever he put his hand to head before I had modeled the movement. This led to some quite hilarious battles of will, with me sitting in one corner of the booth whistling nonchalantly and pretending not to see the baby orang sitting in front of me with one hand glued ludicrously to the top of his head. All the while Cody stared with obvious longing toward his baby bottle. Being by nature quite stubborn, Cody could keep his hand in this position for a considerable time, on one occasion for four minutes. But gradually my nonattention to Cody's activities began to pay off; his spontaneous hand-to-head signs became increasingly rare and lacking in conviction. Within five days the sign had degenerated into no more than brief flappings of the arm, and by the end of the week it was, to use the technical phrase, "extinct." The way was now clear for teaching a second imitatory movement.

Cody's imitation success was indeed thrilling, but even more exciting was his progress in another subject: mouth-breathing. I had kept up this aspect of tuition as a kind of safety net, in case the hand-to-head imitation proved a failure. At first, however, no matter how hard I tried, I just could not persuade Cody to let me close his nose with my fingers. Even the hated "time out" procedures were ineffective. Cody just clamped his mouth tight shut and set about asphyxiating himself into insensibility. His little face suffused with blood until I felt I simply had to let go. After all I couldn't let him kill himself. Or could I?

On the fortieth day of finger-on-nose training, I decided on sink-or-swim tactics. Cody was planted firmly on my lap, and sat looking at me defiantly, his jaw as tightly shut as a bulldog's. He knew what was coming, and I could see he was ready for the fray. I fed him a piece of chocolate to soften him up a little, then slid my fingers over his nostrils. As usual, Cody went wild, squirming and writhing with muffled screams, his arms flailing, but never, *never* opening his mouth. This first session went badly—each time I thought I was about to succeed, Cody managed to twist his head to one side and steal a quick nostrilful of air.

I became increasingly annoyed, partly with Cody, but mostly with myself for forcing the young orang to undergo this torture. Yet I could not give up. If I did, Cody might feel he had bested me, and then all his training could be in jeopardy. I grasped Cody even tighter, locked his head against my knees, and clamped his nose again.

"All right, you bugger," I said between clenched teeth, "die if you want." For several long horrible moments I thought that this was just what Cody intended to do. He fought like a demon to be free, and with such energy it seemed that at any moment he would break loose. It took all my strength to hold him as he bucked and wriggled, all without a single breath, though I could see his little chest working rhythmically as it sought the air his mouth denied. For almost a minute this silent battle continued.

Then, with a grating inhalation like that of a child with whooping cough, Cody took his first mouth breath. I was ecstatic; but no sooner had the little ape inspired than he set his mouth firm again, and the fight began once more until his next breath. Yet again persistence paid off; by Cody's fifteenth mouth breath, we were down to a ten-second interval between mouth breaths, and he was putting up no more than a token resistance. As with "hand-on-head," familiarity was beginning to breed compliance.

Cody was soon breathing through his mouth with ease whenever I blocked his nasal passages. But there was more to it than that. Once again the little ape went further than I had asked *and* in a totally unexpected direction. During the thirty-first day of finger-on-nose training, Cody suddenly began to *vocalize* in a completely unnatural way, giving voice to a deep, gutteral "gruh" sound. It was a sound utterly different from anything an orang—wild or home-raised—should make, and why he did it I shall never know. But I took immediately to rewarding Cody with

Chocolate Buttons whenever he produced it. The frequency of "gruhing" increased (Cody soon grasped the equation "gruh equals extra chocolate"), and I then rewarded only for ever more refined versions of the sound.

The result was that 24 days after "gruhing" began, Cody could give voice to an acceptable "kuh" sound, though he still needed my finger covering his nose. There was a long way to go, but gradually a scheme, a hope, blossomed in my head. I believed there was a short cut to Cody's talking. It was risky, but I thought I could envisage a way of integrating both hand-on-head and "kuh" training so as to teach Cody to produce his first unaided word.

12

Talking was not the only field in which Cody had progressed. While never as unbearably extrovert as a chimpanzee, between the ages of two and ten months Cody had developed into a much more social creature, quite aware of his privileged position in this one-parent family. By the time he had entered his eighth month, he had devised several games that he played with me or any of his regular visitors. His favorite was one we called "Timber!" The game began when the player (or victim—depending on whether one liked to play Timber!) sat on the floor while Cody paced around him on all fours. This went on for some minutes until Cody judged that sufficient tension had built up and he could enter the second, more active, phase of the game.

Cody would suddenly stop his pacing, stand in front of the victim, and fix him with a diabolical stare before rising up to his full height, arms held skyward. He would teeter there for a few seconds, still grinning devilishly, before collapsing on top of the victim with all the force of a felled oak. A quick tussle ensued, in which Cody tried to bite his opponent anywhere he could, and then the four-armed horror would extricate himself and begin pacing again.

Timber! was quite a frightening game, and Cody's rhythmic,

purposeful pacing and his satanic stare seemed to exercise a kind of hypnotic effect on some players. Grown men and women, who could easily have stood up and so ended their torment, never used this method to stop the game, but simply sat it out, all the while begging Cody to stop. This was just what Cody seemed to want, and it only made him the more eager to continue. I usually intervened at this stage to save the human half of the game from nervous collapse.

Timber! was played with friends, but the dubious pleasures of "The Chase" were bestowed upon anyone Cody thought he could terrorize. Each time a child was brought in to see Cody the same procedure was repeated. As soon as he had assessed the newcomer as "challengeable," the young orang would fly straight toward him, mouth open and eyes ablaze. If the stranger retreated, The Chase was on, and it could continue for as long as the pursuer had breath. Cody was very crafty about the whole thing; he was always careful to run just slowly enough to keep on the heels of his terrified guests, but never to overtake and catch them. He played this game with at least four little girls, three boys, and a cocker spaniel.

The only time I saw him outdone was when he tried The Chase with the head keeper's young son, Michael Barclay. For the first two minutes everything went according to plan, at least as far as Cody was concerned. During this time, Michael performed several laps around the outside of the cage screaming loudly while Cody pursued him eagerly, giving Michael a nip on the ankle every now and again to speed him up a bit. Then suddenly a strange expression came over Michael's face. He stopped running, turned round, and landed a neat kick on Cody's chin. It was a good solid clout, and it brought Cody up short. He sat there rubbing his chin thoughtfully and looking at Michael, as if he'd been mortally offended.

Michael held his ground and looked ready to kick again. Cody must have realized this for, with quiet dignity, he scampered back to where I was standing, settled himself below the chair with a tight grip on my trousers, and trained a wary eye on the enemy.

From that time on Cody treated Michael, albeit grudgingly, as an equal.

I am inclined to view both Timber! and The Chase as being experiments by Cody in dominance. Most animals have some sort of hierarchy among members of the same species in which each knows his or her own place. Such dominance hierarchies have an important adaptive value; they are a very good way to organize social interactions. Imagine an army without its chain of command and one can very soon see what a mess humans would be in without some sort of descending order of influence. Lacking the company of his own kind, I believe that Cody used both Timber! and The Chase to establish a similar pecking order with his human companions.

However, not all Cody's games were colored with such ulterior motives. "Wrestle Tickle" was one he seemed to enjoy for its own sake, if frequency was any criterion. Cody played it often, usually once a day, and sometimes nonstop for as long as half an hour. He normally initiated play by lying on his back at my feet. Then he would throw his arms and legs into the air and smile at me mischievously while his little fat belly wobbled to and fro. This was the signal for me to fall on the floor beside him, tickling him wherever I could find an opening, while he tried to fend me off as best he could.

For the most part Cody's ticklish areas corresponded to those of a human: the soles of his feet, sides of his chest, and parts of his shoulders (especially near the collar bone) were all good tickling zones. One area *was* different—Cody would invariably curl up in ecstasy if I so much as touched the inside of his elbows! His furry-red body would heave in paroxysms of delight whenever I found any one of these sensitive spots, and his breath would come in short rasping gasps. It was as close as I ever saw Cody come to human laughter.

After Wrestle Tickle "Fighting Freddy" was Cody's favorite way of passing the time. Freddy the Teddy had been Cody's constant companion since his eighth week of life. Cody always slept with Freddy in his bed, but it was a love-hate sort of relationship. They were growing up together, although for Freddy this was more a case of growing down. Every so often the little ape would take it into his head to beat the hell out of Freddy and, being at heart a bear of straw, Freddy did nothing to stop him. Cody would worry the bear like a terrier with a rat, shaking him bodily or whirling him around his head in a frenzy of viciousness and stopping only to beat him madly against the bed railings. Freddy was also useful for practicing the various biting sallies Cody used when playing his Timber! game. By the end of the experiment poor Freddy was limbless, eyeless, and earless. But Cody would still not go to sleep without him.

As well as being playful, robust, and vocal, Cody was now a thinking orangutan, quite aware of the world around him. He knew all sorts of tricks by which he could manipulate the people about him, everything from crying piteously when hungry to acting all gentle and tender so that he could get his hair brushed. But more than these social subterfuges, Cody at last showed signs of knowing about objects and the way in which objects are related to each other. It had not been like this during the first few months of the infant's life: for example, Cody had been completely ignorant of the fact that an object placed out of sight behind a screen continued to exist. He behaved as if it had vanished into the void, even when he was in the act of reaching for it. For Cody, out of sight meant out of mind.

The infant orang is not alone in this misconception. Human infants, too, lack what psychologists call an object concept. All adults believe that the milk bottle they placed in the refrigerator is still there after they shut the door, even though they can no longer see it. They believe in the permanency of objects, that people and things exist in space and time independently of their own actions. Such ideas may be child's play to the adult, but not to a newborn baby. According to Jean Piaget, the human infant in the first few days of life exists in a universe which is qualitatively different from that recognized by an adult. The baby's world is thought to comprise pictures and sensations which, while they may be recognizable to the infant, have as yet no permanence or spatial organization. A child opening its eyes is rather like an adult switching on a TV set. Images come and go on the screen, and whether they are pleasant or ghastly we have no real control over them, except perhaps to switch off. For the infant human the images his brain projects of the world are at first about as unreal as Batman is to an adult (and perhaps as ghastly). It is only by slow degrees that the child comes to understand that what he sees reflects what is happening "out there" in the world around him. The permanency of objects is very important in this realization.

Babies get into terrible difficulties as they try to be objective about objects. But it is a very serious business. Piaget, in his book *The Origins of Intelligence in the Child,* devised a series of simple tests that broke down the infant's responses when a desired object is hidden behind some sort of masking device, usually a screen or cloth cover. During the first two stages (birth to four and a half months), the human baby makes very little response when the object of his desires is hidden, or even half-hidden. It is as if the object ceases to be. Searching for a half-hidden, but *not* for a completely hidden object indicates a Stage III level of ability "four and a half to eight months) while looking for a completely hidden object (by pulling down the screen that hides it, or removing a covering cloth), is a Stage IV behavior.

Once the infant has learned to find a completely hidden object one might think it understands about an object's permanency. But not so. If, during Stage IV (which occurs between eight and 12 months), one hides an object first under one cloth, and then, while the infant watches, moves the object under a second cloth two or three feet away, the child will be completely nonplussed. Babies of this age will diligently watch these proceedings, then lift the *first* cover, but never think to look beneath the second!

Piaget believes that the infant searches incorrectly, not because he lacks the motor skill or has forgotten the object's whereabouts, but because he has a different set of assumptions about the nature of objects. The human infant is the supreme egoist and behaves as if the existence, or at least the location, of an object depends upon his own actions. It is as if the child believes the act of grasping will recreate the object anew at the site he desires. Not until around the ninth month does this egocentric attitude wane sufficiently for the two-screen test to be accomplished. This is the Stage V level.

The last stage, VI, is achieved when the infant can comprehend an invisible displacement of the object. The experimenter shows the child an object held in his hand, then places it (silently) beneath a cover and holds out his closed hand to the child. While the State V baby will expect to find the object still in the hand, ignoring the covered hiding place, the Stage VI baby will look immediately under the cover. With mastery of this final stage, the infant is said to have attained the object concept.

When first propounded in the early 1950s, Piaget's theories had been studiously ignored. But subsequently scientists realized that his hypotheses had much to offer in the understanding of human development. I was interested to see if Piaget's theories could be applied equally to other animal species. It would obviously add immeasurably to the strength of the Piagetian position if his ideas were found to apply across a wide range of animal life. No one had followed any animal through its early development specifically to test for object concept, a very important aspect of "intelligence." The orangutan, being such a close relative of man, seemed the ideal subject with which to begin this cross-species investigation.

Over a period of months, I had given Cody the same cover and screen tests as Piaget had used on his own children. Luckily Cody was very interested in novel toys, so these could be used as test objects for most of the time. When Cody's interest waned, I used a small bottle full of highly sweetened milk or fruit juice. Being a true orang, in other words greedy, Cody *never* lost interest in this particular object. As the data built up slowly over a series of more than 60 test periods, it became clear that, with a few exceptions, Cody was following exactly the same path to enlightenment as a human infant! Each stage from I to VI followed one upon the other in what seemed to be a preordained series to which, willy-nilly, both human and orang are tied. Piaget's stages *were* valid across the species barrier, at least within the *Hominoi-*

dea, the group to which both apes and man belong. The exceptions to this common pattern were minor, but they were *very* weird.

When Cody was approaching Stage IV he was tested with two screens, A and B. The object, his mini-bottle, was hidden behind screen A for the first three trials, Cody successfully finding and grasping the desired object each time. Then, with the young ape watching carefully (food was involved!), the bottle was hidden behind screen A and then (in full view of Cody), moved behind screen B. As expected by Piagetian theory, Cody immediately and incorrectly pulled down screen A but, instead of noting the absence of the bottle at this location, he proceeded to make a grasping movement on thin air where the object should (to him) have been, and to bring his hand with its invisible object back to his mouth! It was as if Cody knew something I did not. Even stranger behavior was to follow, which made me doubt whether the orangutan was as intelligent as I had hoped. As he progressed with the two-screen test Cody eventually chanced on the correct response. That is, he pulled down the screen B behind which the object had last disappeared. But having removed the correct screen and with the object plainly visible (at least to me!), Cody pulled down screen A and grasped at a phantom bottle behind it! Only then did he turn his attention to the real thing.

At first I could not understand Cody's actions. The theories of Piaget, based on the superlative egocentrism of infancy, would conclude that this was yet another example of the infant believing that an object is under the control of his actions and can be brought magically to the point he desires simply by making a grasping movement. Yet the more I considered this explanation, the more Cartesian it appeared. Like Descartes, Piaget seemed to be postulating a "little man," a sort of naíve "soul," getting inside the infant's skull, brimful with egotism and giving its body commands. This seemed implausible, especially when it occurred to me there might be a simpler explanation for Cody's erratic behavior.

The infant's actions at this stage are very "concrete"; that is, they are limited in number and not easily modifiable. There is a wealth of evidence to support this assumption. If presented with two screens, and an object that is no longer visible, and if allowed to find the object behind screen A and nowhere else on several trials, then the infant (be it human or orang) will perform the action which, in its limited experience, has been found to work previously. That is, it will pull down and grasp at site A. It seems to me therefore that the infant produces an incorrect response because it chooses the action which, in the small sum of its experience, worked previously; and also because it has no better action within its repertoire. This is somewhat different from crediting an infant (of any species), with a magical, self-centered "belief" in its own abilities.

Whatever the "motives" behind Cody's response, the fact remained that this young orang was developing a very similar object concept to that used by a human infant. But what was quite shocking to me, and to every other psychologist I spoke to, was the pace at which Cody developed. The young ape had learned everything much faster than a human child. At first there had been no real difference: Stages I and II were all but identical, while Cody actually took longer to achieve Stage III—five and a quarter months as opposed to the human norm of four and a half. But by Stage VI things had speeded up, Cody achieving mastery of the invisible displacement problem at 11 months—an age when most human babies are still laboring with Stage IV. In the race for the object concept, the orangutan seems to be at least seven months ahead of its human competitor!

This result was a great surprise to everyone. Being quite certain of our own species' superiority in intellectual matters, we had all expected the ape to learn at a much slower rate than man and to stop at some level below us Lords of Creation. In a way, we were equating apes with less bright human children. Being slower than their peers, such school children are unable to take in all the information available to them. Apes, we felt, would have similar trouble and be slow learners, only to a much worse degree. But Cody had completely reversed these ideas; during the first year of life it is the ape that outstrips the human. In terms of learning rate, the human infant is the dunce. But as men are demonstrably and overwhelmingly cleverer than apes just what is happening?

One explanation is that the foundations of an orang's knowledge are laid down a little too fast so that further gains are not possible. By contrast the human species, working at a slower pace, has a much longer time to assimilate the many intricacies of each cognitive level and so has a firmer structure of ideas and concepts on which to build. This idea lends weight to the theory that, during evolution, the slowing down of our psychological development was quite as important as our use of tools, language, and other so-called human abilities. Indeed, these latter accomplishments may well be spin offs, derivatives of the former, more basic development. Taking our time when young might have a great deal to do with the success of our species on this planet.

13

In spring, when Cody was seven months old, he acquired an "uncle," Andrew Lock, a tall, slim and rather laconic researcher from Hull University's psychology department, had heard of my study and telephoned to ask if he might video-record my interactions with Cody. He had been doing similar work on the way in which human mothers and infants communicate by looks, gestures, and wordless sounds, and he was interested to see if cross-species mothering would show similar patterns. It sounded an intriguing subject, and I was delighted to help, especially when Andy offered to record Cody's progress in talking tuition for further analysis. We arranged that Andy should visit every Thursday afternoon for an hour or two's recording session.

These visits became a regular part of Cody's week. It was strange, but he seemed to know when the day for Uncle Andy to arrive had come around. How he did this I cannot think. Perhaps he could tell from the change in routine that always occurred on Thursday. But I would often find him staring out of the window in the direction from which Andy usually came, sometimes two hours before our visitor was due. John MacKinnon believes that the wild orangutan is able to calculate which tree in the forest will be in fruit at a particular time of the year; it may be that Cody used a similar method to calculate his uncle's arrival. Andy was touched by Cody's devotion and was clearly pleased that he had struck up such a good, "meaningful" relationship with the young ape. Perhaps it *was* Andy's personality that Cody found attractive, but I had a sneaking suspicion it was the chocolate bar he always brought along that really sold Cody on his uncle.

One particular Thursday, as usual, Andy arrived punctually in his clapped-out 15-year-old Spitfire. After we had carried the video equipment into the workroom, he immediately crossed over to greet Cody, who was hanging from the playroom side of the dividing weld-mesh door, screaming loudly for attention. Andy was an old-timer at this game now and quickly stood to one side of the door as he approached the little ape. It was as well as he did; Cody's "greeting piddle" surged forth in a shining stream, making a pretty tinkling sound as it hit the floor tiles. Cody reserved this ceremony for those he really liked. I was honored every morning and had got into the habit of seizing the mop as soon as I stepped through the door.

Cody's screams subsided as soon as the chocolate bar was produced, but only for as long as it took to consume the treat. This was about two minutes, not long enough for us to position Andy's gear and make all the necessary adjustments. We were not halfway through the tuning and balancing when the whining and screeching began again, full blast, and by the time we had finished I knew just what to expect from Cody. Almost without thinking, I walked over to the bench, pulled out a clean diaper and then strode purposefully to where the animated fur-rug hung—screaming blue murder—from the top of the door. Our many perambulations across the front of the cage had really frustrated Cody. There were his mama and his uncle, always visible but forever just out of reach. And a frustrated orang is no respecter of diapers. True to form, Cody stank to high heaven. I opened the door, and he swung effortlessly around the metal frame, attaching himself with limpetlike tenacity to my neck and head. The way he pulled at my hair, I was sure I'd be bald by the time this research ended. Having eventually managed to remove his clinging limbs, I laid him on the table and proceeded to change his diaper. Throughout it all Andy's video whirled with a low hum, recording the grisly details. Then began the long job of cleaning and clearing away the broken toys and furniture, the result of the infant's early morning explorations. Cody, for his part, had plans of his own.

He had become increasingly intrigued by the far corner of the workroom. Here was a small baby pram, full of odds and ends. Here were spare clothes. Here was a rubbish sack crammed to bursting with trash, and therefore of great interest to a baby orangutan. But best of all, right in the very corner, and dangling tantalizingly from the bench, was a fascinating complex of wires, long silvery-gray snakes of plastic that curled, twisted, and looped invitingly, ending ultimately at the radio and tape recorder. This was Cody's longed-for goal—the possibilities for destruction were *so* exciting—and each time the opportunity presented itself he would get there fairly fast, but so eager was he to reach this El Dorado that his progress was marked by the noisy tattoo of tiny hands and feet beating rapidly against the floorboards. His miniature drum roll usually gave me time to prevent the annihilation of the electrical equipment.

Cody had just attempted his seventh sally against this particular target, and I was getting more than a little tired. It seemed to me a baby orangutan must be the most annoying creature on God's earth, despite Andy's assurances that he had seen far worse brattishness in humans. Cody's last try for the tape equipment had been his best yet, and I had only just managed to stop him—his hand had closed on the wire leading from the tape and he was beginning to pull it, brown eyes shining with glee and his hairy face wearing the most ferocious grin. I clutched desperately at his hand, unlocked it from the wire and dragged him back to the far side of the room. Cody twittered his disappointment. Why did I always spoil his fun?

"Stay there, you little bugger," I said angrily. "Leave that alone.

It's taboo. Verboten. OK?" He stared up at me uncomprehendingly. I continued sweeping the floor, and Cody began his eighth run.

My irate "No!" pulled him sharply to a halt. He sat on his bum, looked round, and viewed me quizzically. Then came a beaming smile which gave me a privileged view of his special "toy of the day," a soggy wadge of cardboard he had been carrying around in his mouth since early morning. Every now and again he stopped and, like some ruminating cow, reflectively chewed the cud. He looked so silly I couldn't help smiling. "Don't you dare," I said, half laughing.

That was a mistake. Cody obviously interpreted the grin and friendly tone as approval. He bombed away toward the wire at his highest speed of crawl, grabbed it, and pulled before I'd taken so much as a step to prevent him. The tape recorder came sailing downward, hitting him right on the crown of his head, and bouncing from there on to the pile of old clothes. Cody yelped, rubbed his head where he'd been struck, and poked angrily at the tape as if in vengeance. Then he turned and crawled over to me, whining piteously. I picked him up, and when I had made sure he was not seriously hurt I turned my attention to the tape recorder. It was still working perfectly, though how any object could come away unscathed from a collision with Cody's concrete cranium was beyond my ken. Cody snuggled into my chest.

Next it was measuring time. Every two weeks I took 23 separate measurements of Cody's anatomy, lengths, girths, and widths of head, arm, leg, and chest, even the length of his ears. These anthropometric measurements were getting harder to take with each passing month. It was not that Cody objected to the procedure: far from it. He had never been tickled so much, or so well, in his life. During the first few months of his existence, while he was relatively immobile, noting down his growth had been easy. But with Cody's increasing agility had come a new-found sensitivity to anyone touching him. His belly and the soles of his handlike feet were the most sensitive, but in the past few weeks making contact with any part of his body, even his forehead, had been enough to send him into paroxysms of delight. As for the metal callipers I used for some measurements—he only needed a sight of them, and you could see his globular body tense with anticipation of the ecstasy to come. If I so much as touched Cody with the end of this instrument he would curl up, laughing madly, and wriggling and squirming in delight.

An hour and a half later I emerged exhausted from the fray.

Over lunch, Andy expressed his pleasure at being on hand to record the "naughty boy" aspects of ape-human interaction. It seemed to him that Cody was acting very much like a normal human infant of the same age. His comment on my behavior was much less reassuring—I was, he said, quite the typical mother!

My brother-in-science left immediately after lunch having refused the offer of coffee in the Animal Kitchen. I decided to have a cup anyway. As I made my way toward the keeper's yard, my attention was suddenly attracted to a dark figure at the top of a nearby telegraph pole. The sun was directly in my eyes; so I walked a few paces before looking up again. My God, it was Molly, the female chimp! She and her mate, Monty, were supposed to be outside today for the newly-arranged chimpanzees' tea party, but not that far out! The poor keepers, Nina and Dave, were going to have their work cut out trying to entice her down.

Molly was the most lovable of all our chimpanzees, but she had a reputation for teasing, especially when it was time for bed. I had once offered to help in bedding down the apes and discovered that Molly's reputation was not undeserved. The chimps slept in a small bedroom that was separated from their larger playroom by a small gate, closed with a sliding door. Molly would run repeatedly toward this gate, but just as you thought she could do no other than go through it, she would give a huge leap, hit the wall above the doorway, and, turning as she rebounded from the wall, she would tear off to the far side of the play cage.

Monty, who always played the good boy on these occasions, nevertheless thought Molly's pranks very good sport and would stamp and hoot excitedly at her antics, or engage in his favorite pastime of blowing the most obscene and embarrassing raspberries it has ever been my misfortune to hear. Molly would watch you all the time, tempting and teasing until, suddenly tired of the game, she would skip into the cage and settle down for her supper or (never thinking that you could still be angry with her) rub against the bars asking to be scratched.

As I watched her on the telegraph pole, she put out her hand and took hold of one of the wires. If they were not careful, I thought, she would swing along this escape route on to the roof of the pub next door. They'd have one hell of a job to catch her then, I turned the corner and lost Molly behind the tall buildings of the ape house. When I looked again the pole was empty. Quick work, I thought, sauntering through the gate—and then suddenly stopped dead.

Nina and another of the keepers, Chris, were sitting at the foot of the telegraph pole, both sobbing wildly. Monty the chimp was lying at Nina's feet hugging her leg. I looked past the two wailing girls and saw Dave and John Laing staring down at a black motionless shape on the ground. I ran forward. It was Molly, her body grotesquely twisted and one foot badly cut. I bent down and touched her. She was still warm. I grabbed her wrist to feel for the pulse.

"We've already tried," said John. His voice was dull and flat. I looked up. He shrugged. "She's dead. She was electrocuted."

The keepers who had witnessed the accident were reluctant to talk about it. The death of a creature they regarded as virtually human threw them all into a mood of black despair. It was like losing a close friend. Nina, as the keeper who had spent by far the most time with Molly, took it worst of all. She cried and cried, sobbing like a baby for more than an hour and refusing to be comforted.

As a result of this tragedy the entire zoo staff became a lot more safety conscious. Fences were mended, locks checked and the whole security of the zoo tightened. Thinking back, I was horrified at my blasé attitude toward Cody's fascination with the tape recorder. From that time on the area containing my electrical equipment became a restricted zone. I attached bells to each wire so that any movement chimed an early warning and set up a plywood barrier between my ape child and the object of his obsession. I was going to make quite sure we never lost Cody in the same way we lost Molly.

14

Cody's first sound came more easily than I had dared hope. Over the past few months the infant had begun to say his "gruh/kuh" sound faster and faster after I had blocked his nose off with my fingers. His vocalization was gradually shaped to an acceptable "kuh," and so proficient did Cody become with this sound that, at seven months of age, he was able to produce it within three seconds of my closing his nose. Instead of a 20-minute-talking session resulting in ten or 12 "kuhs," I could now get upward of 40 in the same period. It became a big drain on my resources to supply enough Chocolate Buttons!

As well as achieving competence in "kuhing," Cody was now a master of hand-on-head imitation. On most occasions all I had to do was place my hand on my head (I did not even have to say, "Do this!") and Cody would immediately follow suit. According to the rules—that is, the regime by which autistic human children were taught to vocalize—at this stage I should now begin teaching the imitation of other arm and leg movements, then imitation of mouth movements, and finally, sound imitation. But autistic children in the studies I had read had never been as good as the little ape was at "kuhing" when their noses were blocked; it seemed such a waste that Cody's "kuh" should be simply an ornament, play no real part in teaching him to speak. How could I ignore the fact that he was vocalizing and continue to teach only the aping of arm movements? Surely I could utilize the "kuh" sound?

It was a gamble, but I began to think that if I could only combine the two skills, I had taught Cody—"kuh," and hand-to-head—then we might be in business far faster than I had imagined. On the other hand, such a lot could go wrong. No one had ever tried to teach anyone—human or ape—to talk in this way before. Cody might never be able to join these two skills together; or worst of all, he might regress, perhaps losing his "kuh" sound, or even, horror of horrors, refusing to imitate.

This was no idle fear; many of the great apes are prone to what, in human infants, is called negativism—a condition defined by the psychologist as "an attitude of resistance to suggestions coming from other people." Negativism is probably one of the commonest attitudes seen in young children. How many times do we hear the word "won't!" from the lips of children who have been asked by their parents to comply with what, to an adult at least, seems an altogether reasonable request? Daniel Ausubel, an American psychologist, has observed that such behavior often reaches a peak in humans at around the age of two or three years. He links this phenomenon with the withdrawal of baby care: the child must learn not to cry when it falls; not to expect too much kissing and cuddling; it must learn to wait for its food; to keep its face clean; and in sum to submit much more closely to all the social mores involved in becoming an adult. At some point or another, usually around the second or third year, these impositions led to frustration and anger, both of which tend to manifest themselvs as stubbornness or negativism. Such behavior normally disappears as the child grows up, but in the great apes negativism can persist to adulthood. Wolfgang Kohler, who worked with chimpanzees on Tenerife in the Canary Islands from 1913 to 1920 and did pioneering work on the cognitive processes of this species, tells of several instances in his book *The Mentality of Apes*.

Sultan, Kohler's most intelligent chimp, was once required to collect the cores and fruit skins left lying about after the rest of the colony's apes had retired for the night.

> He quickly grasped what was required of him, and did it—but only for two days. On the third day he had to be told every movement to go on; the fourth he had to be ordered from one banana skin to the next, and on the fifth and following days his limbs had to be moved for every movement, seizing, picking up, walking, holding the skins over the basket, letting them drop, and so on, because they (his limbs) stopped dead at whatever place they had come to, or to which they had been led. The animal behaved like a run-down clock . . .

If that was how an adult ape behaved when asked to do something it did not like, I dreaded to think how Cody might respond. If I forced him into actions he disliked it could ruin everything. Should I continue Cody's language according to the operant conditioning book, or gamble with my own combination idea? It was a terrible choice, but there was little time for delay. I chose the gamble.

I now wanted Cody to use *his own* hand to block off his nose, and for him then to say "kuh." Within 20 days Cody was doing just that. At first, I asked Cody to imitate once again: this time, instead of placing my hand on my head, I brought it up to and over my nose, after which I said "kuh" as loudly as I could. Cody's arm was then moved in a similar way ending with his hand covering his nose—I kept his hand in this position until he voiced a "kuh" sound. As soon as this occurred, his hand was released and a Chocolate Button popped into his mouth. Unfortunately Cody was just as averse to blocking his own nose as he had been at first when I did it. Much writhing and struggling ensued, but after five days his aversive actions had virtually disappeared.

The "kuh" sound, however, was at a much quieter level than in preceding stages, and I feared that this change of routine might mean that it would be lost completely. We persevered, and seven days later Cody was not only producing reasonable "kuhs" (I became a great connoisseur of this sound), but was also making most of the hand-to-nose movements himself.

I decided then to try slowly decreasing the amount of pressure I was using to move Cody's hand to his nose. (In operant conditioning parlance I "faded the prompt.") It did not work. As soon as I reduced pressure on his arm Cody stopped completely—it was clear he still did not like blocking off his own nose. It was not for another four days that fading showed any signs of success. However, once Cody realized what was required of him, I found I could fade quite rapidly. By June 25, at an age of 11 months, Cody was producing perfect "kuh" sounds without so much as a touch from me. All I had to do was model movement and sound, and the little ape would follow suit immediately. The following day I faded the visual prompt: just saying "kuh" was now sufficient for Cody to go through the whole of his vocal routine. No parent could have been more proud of his child's first word. At last, after almost six months of heartbreaking effort, I had a baby ape that "talked."

But more than this: on some occasions Cody did not even bother to put his hand on his nose before he let out his sound. This was the first time that an ape had achieved such control of its vocal apparatus. All other apes that I knew of, including the chimpanzee Viki and the orang infant of Dr. Furness had had to place their hands across their noses before they could make their taught sounds. This had led many scientists to believe that apes were incapable of voluntarily closing off their nasal passages by raising the soft palate at the back of the roof of the mouth. According to this theory, it was very unlikely that an ape could be successfully taught language. Cody had proved this hypothesis false, at least in part. He had shown that apes, the orangutans anyway, *could* control their vocal apparatus to a much greater degree than previously thought possible. If they were able to do this, then vocal language—or at any rate a simplified version of it—might not be beyond their reach. Cody's behavior once his first word had been learned made me even more certain that this was true.

Almost as soon as he had mastered "kuh," Cody began to anticipate the training sessions: when I appeared with the milk bottle and Chocolate Buttons, and even before we went inside the talking box, Cody was *using this verbal sign spontaneously—as a request for food!* Unlike the hand-to-head sign, which I had studiously ignored, I now actively encouraged Cody's self-invented request word. It was just what I had hoped would happen, and it augured well for the future.

Also encouraging was the infant's spontaneous use of his sound. Cody actually seemed to enjoy saying "kuh," he would repeat it to himself while he played on his climbing frame or with his toys, or while he just stared pensively out of the window. On one occasion, for no apparent reason, he said "kuh" 11 times in 20 minutes! Such a high incidence of spontaneous sound production could only mean that Cody was interested in the sound for itself and in his new-found ability to produce it voluntarily. Perhaps he was even practicing his word, and comparing it to the remembered model I had taught him.

The way in which Cody was playing with the "kuh" sound was very similar to human behavior when the young of our own species are just beginning to speak. Kicking a new sound around in this way probably helps both ape and child by allowing them to practice the sound and to discover the context(s) in which it is useful. If the infant human learns the word "moo-cow" for cattle and then uses it also for sheep, he will, initially at least, be encouraged and applauded. After all, sheep are quite similar to cows; and he is too young to be taught the finer distinctions. But let him try out the appellation "moo-cow" on the rocking chair or, worse still, on Aunt Dorothy, and he will be gently but firmly dissuaded! Very gradually, the lines between animals, inanimate objects, and humans are carefully drawn by the simple procedure of feeding the infant new words. If Cody was as interested in his sound as he appeared, then the chances were that he would take readily to learning other sounds, especially if each sound gave him some specific reward in return.

I would have been satisfied with Cody's progress had this been all that the young ape had managed to do with "kuh." But Cody seemed intent on wringing the last ounce of use from his single means of vocal communication. How he hit upon the idea, I do not know, but Cody went much further than I had ever expected and began using his sound in a completely original way.

It happened first during a Piagetian "string test." In this test the infant is seated at a table with a desired object in a cup or other receptacle on the far side of the table, out of reach. Attached to the object, and within his reach, is a long string. To solve the problem the young creature must grasp the string and use it to pull the object toward him. Most adult primates, including the orangutan, can accomplish this although the lemurs make heavy weather of it. I was confident that Cody would eventually solve it too, but what interested me was at what age, and in what way, he would come to the solution.

Cody surprised me by obtaining the desired object in an utterly orginal way. He would simply say "kuh" as soon as he saw the object and before I could place it at the farthest end of the string! In the most direct of ways Cody was showing me that he was not just a programmed automaton. He was using his word in a novel way, almost like a tool, to get himself the things he wanted. As I wanted Cody to feel his sounds were of the utmost importance, I naturally had to give him the object immediately he voiced his sound. It appeared that Cody had been coming to certain far-reaching conclusions about the "kuh" sound. He seemed to be working on the assumption that "kuh" was a sort of magic spell that could get him anything he wanted.

Perhaps I should have expected something like this. With human infants, Piaget has described certain "magical movements," which the child seems to believe are important in achieving specific aims. For example, one of Piaget's children, Laurent, was observed to arch his back and shake his whole body in order to rock the toys that hung from the hood of his cot. The child later transferred these movements to other situations: when he wanted to have any interesting spectacle repeated Laurent would arch his back or shake himself violently, apparently in the belief that such ritualized actions could "cause" the things he desired to occur. It looked as if Cody was using his new sound as a sort of vocal equivalent to the human infant's magical movements, reciting his spell whenever he thought it would help. I could see that life with Cody was going to become even more complicated from now on.

15

Once I knew that Cody had really grasped his first sound, I had to give him some time to consolidate it. Gradually I made "kuh" mandatory, not just for the training periods inside the talking booth, but for every bit of food and drink the infant was given throughout the day. I dropped the use of Chocolate Buttons as reward—Cody was now vocalizing for his normal meals, and I could see I might need this superweapon for the next sound I tried to teach. Also, by slow stages, I transferred vocal training from the talking booth to everyday situations.

At first, instead of asking for the "kuh" sound within the closed booth, I opened the door slightly before demanding a vocalization, gradually increasing the gap with each additional correct response. Within a few test sessions Cody was "kuhing" vigorously with the door wide open. Then we sat closer and closer to the threshold, gradually moving outside the closeted environment until we were sitting in the middle of the workroom floor. Once again, this move seemed to make very little impression on Cody's production of "kuh." In all, transfer to the ouside took only 11 days. I even managed to persuade Cody to perform outside his hut and display his prowess to the assembled keepers in the

Animal Kitchen. By this time, I required something new of Cody and his sound; as well as saying "kuh" for sustenance, he now had to say the word whenever he wanted his connecting door opened. Cody learned his new "meaning" of his sound within seven days, a very rapid response.

Almost coincident with this Cody devised another use for "kuh." It was all because he hated to have his diaper changed. As soon as I laid Cody in his prone, diaper-changing position, he would strive with all his might to sit up and then to climb away. This led me to devise a step-by-step approach to diaper changing which I have never seen mentioned in any books on child care. Almost as soon as I had unfastened the safety pin Cody would be pushing himself into a sitting position. I would push him down and remove his soiled diaper, by which time Cody was once again sitting up. This up and down routine recurred while I cleaned him up, positioned a new diaper, and folded and fastened it in place. Instead of taking one or two minutes, changing Cody was a full five-minute job. It was probably the imposed inactivity that he objected to, or perhaps he didn't care about personal hygiene.

Now Cody's new sound gave him the perfect "out." I had only to lay him down for a multitude of "kuhs" to sound loud and clear as a request to be allowed up. Once again I was stuck with the old talking dilemma: should I reward Cody's sound by allowing him to get up and remain in his dirty diaper, or was the vocalization to be ignored, with the risk that Cody might lose interest in his sound? Like it or not, I had to make the choice between a talking and an odoriferous orang. Until Cody had invented this particular use for his sound, I had made it a rule always to reward his "kuh" whenever it was produced. But a ponging pongid was too much—I ignored Cody's protesting "kuhs" during diaper changing and simply hoped for the best as far as talking was concerned. Luckily, this did not seem to impair his vocalization in other circumstances, and Cody retained his sound at peak level.

He could now produce a well-formed, distinct "kuh" during feeding and door-opening sessions, and he continued to voice his sound spontaneously whenever (or so it seemed), the mood took him. The vocalization seemed very well entrenched in Cody's psyche, and I felt the time had come to begin tuition for a second sound. This would be the *experimentum crucis*.

Although Cody's achievements to date were extremely impressive, they still did not tell me if he was capable of using specific sounds for a specific referent. It was one thing to teach Cody a single sound that gave him everything: food, milk, tickling, door-opening, contact comfort. But was it possible for a creature like Cody, with his limited brain size in comparison with our own, to link different sounds to different actions and objects? I was not sure, but Cody's innovative use of his single sound made me quite optimistic that such a skill might not be beyond his capabilities. If this proved to be the case, it would have far-reaching implications for those theorists who concerned themselves with glottogenesis—the origin of language.

Our ancestors, roaming the African savannah some two million years ago, had brains of around 600-700 cc, not much larger than that of the present-day chimpanzee or orangutan. It seems reasonable, therefore, to credit these early hominids with at least the cognitive capacities of the modern great apes. But just when in our long history did language appear? Presumably, when it was needed. This is not as facile an answer as it may first appear. As Adrian Kortlaandt has pointed out, the chimpanzee (and the orang) are primarily fruit pickers and have less to discuss with one another than man, a species that took the hunting path to survival and needed language to engage in cooperative stalking and ambushing strategies. The ultimate origins of language are therefore connected with the social and environmental pressures on the early hominids.

These creatures are thought to have lived in small social groups not so very different from those of some primate species living today. We must assume that the first men spoke a very simple and restricted tongue, unlike any spoken natural language of today. This first-of-all language would not show verb phrases, nor pronouns or adjectives. Only later would the phonetic and grammatical elaborations that we take for granted, have become manifest.

This is by no means the only theory that claims to account for the origins of language. Other postulations range from the belief that language arose through the imitation of animal noises that came eventually to stand for the animal itself (the so-called bow-wow theory) to the belief that language is the outcome of a random biological event or mutation (the chance theory).

Frank B. Livingstone, of the University of Michigan, has even suggested that the australopithecines—long thought of as the direct progenitors of man—came to language by way of singing! Livingstone points out that although it is often said that man is the only primate that can talk, he is also the only member of the order that can sing. As W. H. Thorpe has shown, some nesting birds seem to be able to recognize their own family from among a group of 2,000 individuals, an ability that Livingstone believes would also have been useful to the early hominids. He postulates that the australopithecines, being hunters, would live in bands with definite territories and would use song to defend their land from neighboring groups. The song would also serve to identify which individuals belonged to each "family," so that strangers could be readily identified and expelled from the hunting grounds. Singing is a much simpler system than speech (it has only pitch as a distinguishing feature), and Livingstone believes such songs were extant long before language showed itself on the face of the earth.

The theory has its attractive aspects, but like singing in apes and the gestural origin of language, it still does not tell us at what point in our evolution spoken languages became a possibility. One of the prerequisites for language is the ability to link discrete sounds to different actions of objects—an ability that would have been vital in the hunting and foraging from which our race is thought to have developed. If I could show that Cody had this capacity, then it would have to be admitted that early hominids with a similar brain size to his had at least the potential for a simple proto-language.

We would also have to change our attitudes toward the linguistic potential of the great apes. Their abilities, while in no way equaling *Homo sapiens* in terms of sophistication and complexity, would at least make these cousins of ours appear far closer to us mentally.

We might then think twice before using such sentient creatures in the more horrible biomedical and psychological experiments. Chimpanzees have been reared in total isolation, innoculated with syphilis, and placed in zero gravity so that they inflated like living balloons. There was even talk of using anthropoid apes as "models" in motor-accident experiments. The animals were to be strapped into the front seats of various types of car, which were then to be crashed at high speed. Perhaps, when we know just how intelligent the great apes are, there will be an end to such morally indefensible experimentation.

What was certain was that our high-flown opinions of ourselves would suffer; in particular, the concept of man the speaker would take a thorough beating.

I thought it advisable that Cody's second sound should differ as much as possible from the first, within the limits of ease of training. This would allow him to discriminate more easily between them, and therefore facilitate learning. The sound I chose as Cody's new word was "puh." It was almost the complete opposite of "kuh": "puh" was a "stop" sound produced by closing off the lips as air was expelled, as opposed to "kuh," which was sounded from the throat; it was unvoiced and therefore required no involvement of the larynx, as "kuh" did; and the early stages of "puh"—a pout—could be visually modeled, which was not the case for "kuh." Another plus was that pouting was already part of Cody's behavioral repertoire.

As well as making the sound as different as possible, I also needed to manipulate the training environment so that when we began "puh" tuition Cody did not feel so at home that he immediately began "kuhing." I was loath to scrap the talking booth where Cody had learned his first word. It had proved invaluable in shielding him from all those distracting objects and events in the outside world. The booth would have to be modified, but I wanted to do it in such a way as to maintain its desirable aspect. I decided on a change in the color scheme. The door of the booth was made reversible and painted the most hideous buttercup yellow. During "puh" training the colored side faced inwards, and I sat with Cody so that I was framed by this startling, yellow-hued backdrop. I also took pains to make myself a "discrepant stimulus" during the first days of training; I wore a saffron mountaineer's cagoule, hooded, and reaching almost to my ankles. I certainly hoped it would work. I often felt more than slightly ridiculous as I sat, incarcerated in a yellow box, looking like a giant pear drop and saying nothing but "puh" to a hairy midget.

Surprisingly, it all worked quite well. After four days of my manipulating his mouth into the required shape and then shoveling in his food reward, Cody had learned to pout when I pouted. Then 14 days later, he produced a quite superb raspberry, one of which Monty the chimp would have been justifiably proud. This was only slightly unfortunate in that Cody happened to have his mouth full of gruel at the time. I was peering down at him earnestly and pouting and "puhing" madly in the hope that he would do the same. He did, and as my face was only four inches from his, the resulting porridge storm covered my face with sticky, salivary oat flakes. This set the pattern for the next two days; I pouted and said "pugh," and Cody raspberried my physiognomy with oatmeal fragments. I had tried changing to liquids in hopes of avoiding the worst of the storm: but if anything is worse than a deluge of gruel it is a shower of hot sweetened milk. In a perverted way, I felt rather proud. Not every researcher is dedicated enough to undergo the indignity of being sprayed with cereal and honey 20 times a session, morning, noon, and night. But to my mind, it was a small price to pay. Cody had learned his second word! Then suddenly, everything went horribly wrong.

The very next time I asked Cody for a "kuh" sound, he responded with several excellent "puhs." No matter what I tried, the infant ape refused to say his first word, but simply kept on repeating the newer acquisition in his vocabulary. It was the same all day—Cody had apparently decided to substitute "puh" for "kuh" in all situations. This was just what I had feared most of all—as soon as "puh" was learned, "kuh" had been lost. I spent the next few days in almost nonstop reteaching of "kuh," and finally, six days after Cody had lost his "kuh," it came back.

But then once again—disaster! No sooner had he relearned "kuh" than his "puh" disappeared. Now it was "kuh, kuh, kuh" with never a "puh" to be heard. I became very depressed at this turn of events. Cody's behavior seemed to show that he was not capable of learning two sounds at the same time, let alone relating different sounds to specific objects. I tried to rationalize this apparent defeat. At least, I told myself, I had discovered that orangs cannot learn to use sounds as "names" for specific objects and actions. But this was cold comfort.

There were now only two courses of action open: give up, or try again. It was mainly because I loathed the idea of giving up, rather than because of any faith I had in Cody, that I did try again. In six days, his "puh," had returned and, wonder of wonders, except for occasional "slips of the tongue," he was also making the "kuh" sound—and in the correct contexts. I was ecstatic: the really crucial part of the experiment was a success, Cody did seem capable of linking "words" to specific contexts. The next stage was to find out how many words he could cope with.

In the face of this triumph, one question still puzzled me—why had this seesaw effect between "kuh" and "puh" come about? And would it happen with each new word Cody learned? The more I thought about it, the less pessimistic I felt. It seemed to me that, far from being an ill omen, Cody's alternate loss of each of his sounds spoke well for his emerging skill at relating sounds to objects. The most plausible explanation for this anomalous behavior is that the acquisition of a second sound occasioned the first real challenge to Cody's hitherto omnipotent "magic spell"—"kuh." Until he had learned "puh" Cody's "kuh" sound had been his giver-of-all-good-things; *anything* he desired was only a "kuh" away. I had been at pains to ensure that whenever Cody "spoke" he got whatever he wanted, and the little ape had quickly become accustomed to this happy state of affairs.

But learning his second sound—"puh"—must have started quite

a conflict in Cody's young mind. "Kuh" was now no longer all-powerful and, because at that time there were more "puh" sessions than "kuh" sessions, the first sound was dropped in favor of "puh" as Cody's "spell." It required several days of training and retraining for Cody to "realize" that both "kuh" and "puh" were required on *specific occasions.* When that happened Cody had at last crossed the threshold. He finally understood the concept that words mean something and were not simply sounds to be played with.

That this interpretation was close to the truth became clear as Cody mastered other words. Each time he lost the previously learned sounds. But, since he was now accustomed to the idea that a specific sound stood for a certain activity/object, Cody's "word loss" never lasted longer than half a day. It seems hard to believe now, but each one of us went through a similarly puzzling time as children. Many authorities have followed the way in which children learn what sound "stands for" what object and found that human infants invariably experience conflict when their vocabulary increases and they must learn increasingly more specific referents for each of their words. For example, when first learned, the word "doggie" is used, not simply for members of the *Canidae,* but also for cats, horses, cows, even the rhinoceroses! Initially, the parents of the child accept this rather strange definition of doggie, just as I let Cody's "kuh" have far-reaching connotations, but gradually, they begin to demand that the child be more precise, that he use the correct word for the job. This is just what I had begun to do with Cody, and it was this that had produced confusion. In a similar way, human infants often misclassify "doggies" and "pussies" until they "get the hang" of the more restricted meanings of these words.

So Cody now had two words in his vocabulary. Apart from my fiancée, Liz, and my parents, whom I had telephoned on the night of the breakthrough, no one else yet knew of Cody's new-found ability. There was a very good reason for this secrecy. In a short while, Liz and I were to be married at her parents' home in

Barbados. When we returned, I certainly did not intend to live in a tiny trailer at Flamingo Park Zoo while my new wife spent her time studying in Durham. But Cody, I knew, was only at the very start of his talking ability, and it seemed a great pity to leave him just at the time he seemed ready to "flower" linguistically. There was only one way out of this impasse. Cody had to come with me to Durham.

This was easier said than done. But I surmised that if I could get two words out of him and then show the directors of the park his achievements, there might be a fifty-fifty chance of them agreeing to my taking the infant with me. Where he would stay, how we would transfer him, I had no idea—all that would come later. The first requirement was the two words; and now we had them. I went immediately to the park director, Mr. Jackson, and told him the news. He was elated. It would, he assured me, be a sensation, the perfect crowd puller. They certainly could do with that at this time of the year. Then he frowned. Was I sure? Could Cody really say these sounds, or was I deluding myself? He didn't want to make this announcement to the world and then be made a fool of. Could I guarantee Cody's performance?

I was ready for this one, and suggested that he and the rest of the directors should come over to Cody's cage where I could give a demonstration of his "talking." Jackson agreed immediately, and I left his office feeling very pleased with myself. All the big guns of the zoo administration would be there. It was the perfect opportunity to put, as forcefully as I could, my arguments as to why Cody *must* be allowed to stay with me in Durham for an extended period.

An hour later I glanced at my watch. 11:45 A.M. I had arranged the performance for 11:30. They should be here by now. For his part, Cody was unconcerned, busily doing his own thing with Freddy the teddy bear, whom he was now attempting to decapitate.

"Do you think they'll come?" I said uneasily to the snuffling bundle of orang and toy—it was hard to see which was which. Cody pulled off the teddy's right ear with an elaborate flourish, sat up, stared happily at his trophy, and then dived into the attack once again.

As his teeth sank into the defenseless teddy's nose, I spied the four directors marching abreast around the corner of the enclosure and making their way in our direction. I carefully extricated Cody from what remained of the hapless Freddy and, dusting the straw filling from his fur, told him to do the best he could. That went for me, too. I had also to give a good account of myself and, providing Cody worked well, I felt I could. The night before I had labored long into the small hours on a three-page, closely reasoned proposal entitled *The Necessity of Taking Cody to Durham for Talking Tuition*. This outlined the impeccable logic behind my wish to remove Cody from the zoo for at least a three-month period. All I had to do was keep to this plan when explaining my requirements, and I ought to be home and dry. Nevertheless my palms were sweating profusely when I opened the door to admit the four men who would decide Cody's, and my own, future within the next half an hour.

They arranged themselves in a rough semicircle around me as I got out the prepared cereal and explained what (I hoped) Cody was about to do. A brief embarrassed silence followed, during which Cody nearly ruined everything by almost knocking the food plate from my hand. The director spoke up.

"OK, let's see him, or rather hear him, then," he said shortly, and we were off.

Sitting down, I raised the first spoonful of food from the plate and looked anxiously at Cody. As usual he stared longingly at the food, but I was still extremely nervous. Would he do it? I opened my mouth to model the sound, but before I could utter a peep, Cody screwed up his eyes, opened his mouth and with a look of the utmost concentration voiced an absolutely perfect "puh."

With shaking hand and spoon I ladled in Cody's reward and then looked up. All three directors were staring silently at Cody, jaws slack with shock, as if they had just seen a UFO land.

"Amazing," said Mr. Boyd. "Amazing," he repeated, still gaping in wonder at Cody, who was already doing his best to reach the second spoonful of food. A second "puh" followed, then seven in rapid succession until the food bowl was empty. By this time the directors, remembering just how important they were, had regained some of their composure. However, they were still very much impressed.

"That's his first word," I said happily, "it means food." I felt like some proud parent whose first-born has just reached an important landmark in its development. "He has a second sound which means drink. This was much harder to teach, but he's certainly got it now."

They looked suitably impressed and requested a demonstration at once. I poured some baby milk into a mug, sat down again, held up the mug and fixing Cody with a steely stare, said "kuh."

Cody looked at me gravely in the eye, modeled a perfect pout and blew a beautiful Bronx cheer. Oh God, I thought dismally, he's regressed again. Wiping the spray from my face, I tried again.

"No Cody, kuh!"

Again the response was a superb raspberry.

"Kuh!"

"Ppppaaarrrppp!"

"Kuh!"

"PPAARRPP!"

"Kuh," I screamed.

This time Cody took in an enormous lungful of air, made his lips a trumpet, and with wide staring eyes uttered a blurt of blurts that reverberated all around the room.

"No, you imbecile." I shook him angrily. "Say KUH!"

Contorting his face into ever more pronounced pouts, his little hands clenched tightly in frustration, Cody continued his raspberries while, now oblivious of the directors, I screamed strident "kuhs" at the pouting pongid.

In desperation, I tapped his mouth with the edge of the cup. Cody parted his lips to take the proffered cup, and as he did so I said "kuh" for the fifteenth time.

The open mouth must somehow have cued Cody to what was desired of him, for he produced a fine "kuh," loud and well formed. After this, to my great relief, there was no more trouble. He managed fourteen good "kuhs" before the milk was exhausted. Suddenly I remembered the directors and, with some trepidation, looked up.

"Great," Bill Boyd exclaimed. "Great, just what we need for the summer season."

"Definitely," David Cooke agreed. "We can bill it as the start of something big—a talking ape! If it comes off, fine. If not, then we'll still have got a goodly load of publicity out of it."

I sat, tense and worried, waiting for a chance to ask my all-important question, while they tossed several ideas about between themselves, how to bill it, when, which angles to push, etc. They were still debating the finer points as, Cody and myself forgotten, they turned and made to leave. I could see I'd just have to jump in with both feet.

"What I've been wondering," I began in a loud voice, "is can I take Cody to Durham with me at the end of the season?" I pushed forward my proposal. Bill Boyd took it and looked down at the title.

"You want to take him to Durham?" he said quietly.

"Yes," I began, intent on belatedly convincing my audience, "for three months. If you read the typescript, especially subheading b . . ."

But Mr. Boyd was already looking around at his colleagues. No one spoke. An eyebrow raised here, a cocking of the head there, was all the indication they gave of their opinions. It meant nothing to me—were they agreeing or rejecting my request? But Mr. Boyd seemed to understand these barely perceptible movements perfectly.

"Right then," he said, turning to face me, "we don't see why not." He pulled at his nose. "Yes, that should be OK." They turned to leave again.

"I can take him?" I could hardly believe it.

"Yes," he answered, "but only at the end of the season. Before that we've got some hard selling to do!"

16

I surveyed Cody's room for any article I might have forgotten. I could see nothing else I should take, but just in case, I checked the small plastic bag I carried in my left hand: three overalls, two diapers, a baby bottle quarter-filled with milk powder and sugar, a banana, apple, two plastic pants, and a spare safety pin. That seemed to be everything. I was becoming quite proficient at traveling with a baby orang round my neck.

John Laing beeped his horn impatiently, and with Cody dangling from my left shoulder, warmly accoutred in a blue woolly sweater, I slammed the door and climbed into the front seat of the curator's aging Mini. John pulled the starter, the car growled twice, fired, and for the third time in four days we were on our way to "star" on TV. Since the zoo's PR man had got to work ten days before, Cody had become an instant celebrity. There always seemed to be photographers calling at the hut for shots, editors ringing up for interviews over the phone, and, most time consuming of all, television appearances.

The directors were delighted. As far as they were concerned the more publicity the better. I was a little less enthusiastic. Naturally I was pleased that people were taking an interest in my work, but the very number of commitments which the zoo had arranged was preventing me from doing any serious research. I was torn between the urge to continue my work as I had been doing for the past 12 months and my desire to help the zoo. After all, they had given me the chance to study Cody, and I felt I owed them something. However, it would have to stop soon or my research schedule would begin to fall apart. I planned on getting more sounds from Cody, and I wasn't going to let anything jeopardize that.

Our first two journeys to TV stations had been fairly local—Manchester and Leeds, both of them under a hundred miles. Whereas for these journeys a car had been sufficient, this third interview was in London. Cody and I would have to use the train, and I dreaded the prospect. It would take at least 11 hours for the round trip, which was tiring enough, but having to restrain Cody's nonstop explorations was going to make it a nightmare experience. There never seemed to be enough hours in the day for this 18-pound bundle of mischief—he never slept until around 11 at night if someone was there to play with him, and to spend more than three hours with the hairy horror was gambling with insanity.

Cody was chewing the seat belt reflectively as we sped along the road to York, and I was thankful that the car journey didn't seem as if it was going to be as "eventful" as the first two. The infant was by now rather blasé about cars. From the start, he had seemed totally uninterested in the fields and farmhouses that flashed by, but what had proved a great draw, on the first of our tours, was the car's interior. It was a long time since Cody had been presented with such an alluring array of levers, knobs, panels, all of which just begged to be investigated. He set to work with a will. His front teeth left twin rail marks on any chromium surface within range, window frames were bitten, and much time spent chewing the fascia and upholstery. Then he discovered the map pocket on the nearside door, and, teeth glinting, homed in on it head first. Every now and again, just for fun, he'd climb back up and try to bite my fingers.

Things were a bit quieter this time, though not much. When the windshield wipers were pressed into service during a sudden shower, Cody flung himself repeatedly at the glass, trying vainly to get a grip on these tempting black rods that flashed backward and forward with such tantalizing regularity. When that failed, he tried a "kuh." All this did not exactly endear Cody to the curator. And his sudden pulling of John's arm—which ran us three times onto the shoulder—finally put paid to Cody's explorations. We were banished to the back seat.

By the time we had reached York, since it was the holiday period, both the roads and pavements were congested with cars, trailers, and tourists, the latter eager to see the sights of this historic city. With all the traffic lights there were many hold-ups, and I was very apprehensive in case people should spot Cody and crowd around the car, pointing and staring. To my surprise this never happened. People's usual reaction was to stare into the car, seemingly directly at the young ape, and then look away with not so much as a glimmer of recognition. Cody seemed to be invisible. It was almost as if, never expecting to see a baby orangutan in the car, the brains of most people simply edited out the hairy apparition that their eyes reported before it was sent for conscious consumption. I don't know how many other unusual phenomena we miss because of our subconscious censor, but I have a feeling it is many more than we could care to admit.

Strangely enough, the one person who did notice Cody was more than a little inebriated. As we waited against yet another red light, a gray-haired and slightly scuffy old man in a long military overcoat came staggering down the steps of a public house in an obviously cheerful frame of mind. He glanced into the car and looked briefly toward Cody who, with his face pressed hard against the window, returned an unwavering stare. Perhaps the alcohol, lapping at his brain cells, had lulled his internal editor to sleep, for the old man stiffened and, goggle-eyed, slowly backed up to the pub's entrance, horrified. Then he turned about, set his now pallid face straight ahead and walked off mechanically down the road, steadfastly denying himself a backward glance. You could hear him thinking: "I'll never touch another drop."

Standing in the line for my ticket at York railway station, I became the target of many curious stares and questions. If this was a foretaste of my journey to London I decided I'd either have to give up now or at least travel first class.

Ignoring the comments of the ticket collector as to what my wife must look like, I passed through the barrier and headed up the platform toward the first class carriage. The head steward was helping various well-dressed dignitaries aboard when I hurried up, breathless and red-faced, a bag in each hand and dangling Cody like a hairy knapsack from my shoulders. He looked me up and down coldly as I held out my ticket. I was quite obviously at the wrong end of the train. He looked first at the ticket, then at Cody, and I noticed a bushy eyebrow rise involuntarily in surprise. But he was the head steward and determined not to lose his customary aplomb. "This way, Sir."

Being a student, all my train journeys had been by second class (when I could afford it), and the contrast between the overcrowded and untidy carriages I was used to and the beautiful hotel-on-wheels into which Cody and I now stepped quite took my breath away. We were directed to a plush seat, before which was a small table covered with a sparkling white cloth and illuminated by a personal reading light. As the train pulled out, tea (at no extra cost) was served along with a variety of cakes. And praise be for the reticence of the English upper classes! Here were no clamoring crowds, no pushing and poking—as I passed along the carriage to my seat the reaction was the same on three separate occasions. A bowler-hatted head would glance up quickly from behind the *Financial Times*, and then hurriedly bury itself again amid the stocks and shares.

At Doncaster—the final stop before London—a middle-aged, immaculately dressed gentleman boarded the train with his son, a smiling tousle-haired youngster. The father refused to acknowledge that anything was other than it should be, and it amazed me to see him go through the motions of requesting an aperitif, drinking it down, and then ordering lunch and eating it, without casting so much as a glance at the hairy apparition that hung silhouetted against the carriage window, swinging by one hand from the curtains.

Between times Cody amused himself by dismembering the table lamp, devouring vast numbers of sugar cubes, and taking several chunks out of a piece of stale bread the steward had offered him. He chewed the table cloth, finally pulling it so hard that cups, saucers and cakes were sent flying across the carriage. The stewards endured the mess, the cacophony of noise, and the breakages with great good humor—Cody was treated like a real VIP, and they seemed quite proud that their train was the first in the country to carry an orangutan in such style. They might not have been so pleased had they seen the two gashes in the fine leather of the seat, the result of an attack by Cody's canines. Each time someone came over for a chat I had to lean across, smiling innocently, and place my hand on the armrest, screening the incriminating holes.

When we at last arrived at King's Cross I was exhausted; the last thing I wanted to do was sit in front of TV cameras and pretend to enjoy it all. I half hoped that the cab driver would not be there, but as I neared the gates a chubby, ruddy-faced young man stepped forward and directed me to the waiting limousine.

The pavement fronting the dark-tinted windows of the Thames Television building was crowded with cameramen as we pulled up. After a ten-minute photo session, we made our way into the building for some welcome refreshments. As we were drinking our coffee, John Stapleton, who was to interview us, joined the group. We went through the sort of questions he would ask me on the program: why was I doing this sort of research? How was Cody responding? Was he a unique orang?

Cody did not seem in the least worried by the change of location, though he did stay very close to me—his usual reaction when confronted with a perplexing or slightly frightening situation. On more threatening occasions Cody hitched a ride on my back. On this occasion the little horror hung to my calf most of the time, making infrequent forays out to other people's feet where he tried his best to chew shoes, ruin tights, and perform all sorts of other jolly japes.

Coffee finished, we had our noses powdered by the make-up lady in order to prevent glare during the interview. After trying desperately to consume the powder puff, Cody got quite a bit of this particular cosmetic up his nose and spent the next two minutes coughing and spluttering on the floor.

We were on the point of leaving the room when a telltale pitter-patter informed me that Cody was relieving himself. Apologizing profusely to everyone for the delay, I sat down again, pulled out diaper, tissues, and powder, and quickly set about making our "star" presentable. John Stapleton was in a state of advanced agitation. "Hurry, hurry," he called, "you're on in two minutes!"

I must have broken all diaper-changing records, for we rushed into the studio with 45 seconds to spare, and took up our positions to the left of camera one. The room was all hustle-bustle, with numerous technicians scurrying from location to location, controlled by three equally large cameramen. The floor manager wandered about, ordering everyone to their appointed places, and speaking by walkie-talkie to the Great God, the producer, who sat enthroned among his shadowy minions in a raised and darkened room to my right.

I was cued in and, grasping a newly made bottle in one hand and Cody in the other, walked out under the bright lights, to the seat provided.

The interviewer squirmed around in his chair as he asked his questions. He seemed to be enjoying it all in a rather self-conscious sort of way. He had the look of a man used to being very grave and serious, who suddenly found himself in the middle of something so ridiculously funny that he was all but bursting with the incongruity of his position. Two or three times, after Cody had said his raspberrylike "puh," I thought he would end his questions by falling forward and giggling uncontrollably, but somehow he managed to hold out. The cameramen were also enjoying the fun and not having to be so restrained: there were numerous sniggerings and guffaws—all carefully out of microphone range—throughout the program.

John was so taken with the voluble anthropoid on my knee that he asked if he might hold him for a while. Perhaps he wished to hold the world's only talking orang, I don't know, but I had no objection and handed the reluctant infant over.

"Can I have the bottle too?" John held out his hand.

This was a different kettle of fish; feeding Cody was a dangerous game, an art in itself. But the cameras were whirring and John was waiting, how could I refuse? Against my better judgment, the bottle went across to John. He put the bottle to Cody's lips, and the little ape seized it avidly, drinking deep. Milk was his favorite food, much preferred to any solid. The interviewer was quite entranced; he looked up, smiling first to the camera, then to me. "Is he always this well-behaved?" he asked me pleasantly.

"Not always." I could see he was having difficulty removing the bottle from Cody's mouth: the baby ape was clinging on stubbornly against the loss of his milk supply. Seeing John trying to pull the nipple from his lips by force I leaned forward in an attempt to prevent the inevitable. "It's best not to . . . "

My voice trailed off as the bottle suddenly came free, but with the nipple still clamped firmly in Cody's mouth. Milk gushed in a creamy white stream from the open top of the bottle on to the interviewer's pin-striped waistcoat. I grabbed Cody and pulled him back on to my lap. John was brushing furiously at his drenched clothing with both hands.

"Yes," I continued, "sometimes he's a bit stroppy."

"Umm, er . . . yes," John was obviously a little flustered. "I can quite believe it." Regaining his composure somewhat, he smiled toward me. "Well, er . . . that was most informative. Thank you, Mr. Laidler. And you too, Cody."

John might have thought the interview finished, but Cody had other ideas. He had consumed no more than a third of his bottle and, to Cody's mind, that was certainly not enough. The little ape wanted more, and he thought he knew how to get it. At home all he had to do was say his "word" and milk would arrive *tout de suite*. So Cody gave vent to a series of "puh" sounds to "command" the food into his presence. As there was no milk at hand, and as I had been asked beforehand to remain seated while John brought the program to a close (we were the final topic that evening), Cody was doomed to disappointment.

After a short puzzled silence, Cody began again, his eyes protruding with the effort, his long mobile lips elongating into a fleshy trumpet and his cheeks inflating like two brown balloons. Cody was not to be stopped, and his "puhs" gradually degenerated into a stream of loud obscene raspberries as the harassed interviewer—still dripping milk—attempted to wind up the program. The theme tune began, and as the "On the Air" signs slowly dimmed, the whole studio collapsed into uncontrollable laughter.

I will always regret having missed the showing of Cody's interview (the producer decided to show it in all its terrible totality); but by the time the program was broadcast, Cody and I were already on our way back to York.

17

The phone was ringing. Cody, dressed in a dark pin-striped suit, collar, and tie, walked from the window, through which he had been gazing absent-mindedly, bent over the desk, and picked up the receiver. "Hello," the voice was perfect BBC English. "This is Cody speaking."

But the phone hadn't stopped. It went on ringing incessantly. Why was that? Why didn't it stop?

Slowly the image of Cody dissolved, to be replaced by the much harsher reality of an English dawn. The strident ring of the telephone—my alarm clock—persisted. Oh God, why had I put it out of arm's reach?

I have always been two different people: the resolute dedicated evening person who, full of business and sincerely meaning to rise early, sets the alarm for 6:30 and leaps determinedly into bed; and the bleary-eyed, weak-willed worm, peering sleepily from beneath the blankets at the cold bleak world of the morning, who finds it a great effort even to turn over, let alone get out of bed.

But today was different. My dream was pure symbolism. It reflected my hopes far more plainly than I ever would have dared admit. Intuitively, if not scientifically, I knew that Cody had it in him—that with careful tuition his "vocabulary" could be extended indefinitely until he could communicate. It might not be the perfect Oxbridge accent of my dream, but his skill would certainly be sufficient to make requests, comment on objects, and perhaps even to ask or answer simple questions. The reason for all this excitement was that Cody had learned his third word, and it had taken him only two weeks! For the nth time I went over Cody's progress to date—anyway it kept me in bed that much longer.

The new sound I had taught Cody was "fuh." The young ape took readily to the mouth shape required (top teeth over bottom lip), and mastered this position within five days. Even allowing for the increased motivation engendered by my new secret weapon—chocolate fudge—this was quick work indeed! Mouth breathing was even easier; I found that my initial plan of closing Cody's nose to encourage him to blow air through the newly learned mouth shape became redundant in three short days. Cody had grasped the idea of making a sound in this way far faster than I could have hoped.

All the signs indicated that my ape child was gradually mastering the art of learning sounds, that he was developing a "learning set" for his tuition. This phenomenon was first described by Harry H. Harlow in his work with monkeys. Harlow gave his monkeys a long series of discrimination tests, one of which involved choosing between two differently shaped objects. One object was "correct" and resulted in a food reward, the other was "incorrect" and meant no food. After the monkey had chosen the "correct" shape on several trials, Harlow switched the test around so that the "correct" shape became "incorrect" and vice versa. He then measured the number of trials required for the monkey to solve the problem (i.e. to switch his choice to the previously incorrect shape).

When he looked at his results, he found that monkeys did increasingly well as the number of test sessions increased. Instead of persisting in their choice of a previously correct shape (as was the case when testing began), the monkeys now swapped sides as soon as the correct shape became incorrect. It seemed they were not only learning the right shape to choose in each session, but also the principle behind the whole test series. Put verbally, what they had discovered was "when the shape that usually gives food does not, change to the other shape!"

With Cody, this showed itself in the way in which, as training proceeded, he had *learned to learn* what was required of him in mastering mouth positions and expelling air through his mouth. The more Cody learned the faster and more efficiently he would learn. If all these heady predictions were true, I would hear Cody's next sound (I was already planning his fourth) in an even shorter time than three weeks.

But first I had to make the three sounds Cody had mastered more stable. Up to now these words had slowly evolved in meaning as tuition progressed. "Kuh" had begun its life meaning "give me" or some similar phrase. Later, "kuh," had come to mean "give me" in the outside world, but Cody had then to learn that, in the training box, it was "puh" that also meant "give me." This was unavoidable at the beginning of the training, but it must have been terribly confusing to Cody and may even have slowed down his rate of progress. Now he had a "fuh" sound, and such chopping and changing would have to stop.

I decided to make "kuh" stand always for milk or other liquid nourishment, and "fuh" for solid food. "Puh" would be different; I wanted to divorce it completely from any dietary context. Cody desired contact comfort just as much as nutrition, so "puh" would become his "pick me up" request. Any other sounds he learned would be given other meanings.

Cody learned these new meanings quickly and soon assailed me with requests to satisfy his every whim. It must have been marvelous for him: whereas before all Cody could do whenever something was amiss was cry and scream, now he could tell me what he required pretty well exactly. Things were not so marvelous for me. Because I wanted Cody to feel that these sounds were important and useful (and therefore profitable to learn), I felt duty bound to fulfill his every request. The "kuhs" and "fuhs" tended to come in bursts or clusters as the time approached for each of Cody's four meals, but "pick me up" was his most insistent and never-ending demand. And a very tiring one it was.

Cody now weighed a healthy 21 pounds and, although more than a year old, he was still something of a "mother's boy." There was nothing he enjoyed more than to play the baby, to be picked up and cuddled and to spend all morning or afternoon riding on my hip and reaching out his over-long arms to grasp at any object that took his fancy. That usually included my pen when I was writing notes, the faucet as I washed diapers (or soap—Cody considered soap a gastronomic delight), or a kettle of boiling water when I was making his porridge. Until he had learned his "pick me up" sound, I had taken every opportunity not to carry my infant ape—it was safer to let him hang on to one leg and to drag him around the room like a ball and chain when I did my chores. But now, I would no sooner put Cody down than he would be standing in front of me, arms upraised and calling "pick me up" over and over again. At times like those I even regretted teaching him his words.

There were other pressing problems, though these were of a far more pleasant nature. I was going to have to leave Cody for a full 28 days in order to get married and go on my honeymoon. My fiancée. Liz, had been born and raised in Barbados, and like all Bajans she had a strong feeling of family and neighborhood. Both she and her parents wanted our marriage to take place at their own parish church. Afterward, we planned to fly to the beautiful island of St. Vincent for our honeymoon. Although I was looking forward to all Caribbean sun and sand, I was still very worried about my research. I was unsure of the effects that such a long absence might have on Cody's words.

One thing I made sure of was that while I was still with Cody we both worked as hard as we could. The last seven days before I left were one long frenetic training session, though Cody was not always the pupil! I did tutor Cody for five to six hours every day, but the rest of the time was spent in training the keepers, who were to care for Cody while I was away, in the intricacies of talking tuition. It was not really a part of their job, and they certainly were not going to be paid for their time and trouble, but everyone who had seen Cody "talking" seemed to have become committed to furthering his linguistic education. Given the short time I had to train the keepers, there was little chance of Cody progressing while I was in the Caribbean, but with the help of the zoo staff I hoped he would not regress either.

Apart from his words, I simply did not know how Cody would react emotionally to being "abandoned." Except for a week's illness, I had never been away from him for more than two out of every 14 days. And the keepers had told me that even during these brief periods Cody became very "sulky." He would refuse all of his solid food, most of his milk, and would usually decrease in weight by anything up to three pounds. Cody's reaction was virtually identical to the *anaclitic depression* documented by R. A. Spitz as early as 1948. The following description refers to a human child, but the similarities with Cody are striking!

> The picture changes on failure of the mother to return. He becomes quiescent, does not look up when adults enter the room, does not play and does not grasp at objects. Along with passivity and dejection he develops eating difficulties, loses weight, and shows sleep disturbances. In general the level of development does not proceed normally or even drops.

"Mama" did not feel too happy about parting either—I was only just beginning to realize how attached I had become to this little ball of mischief. Cody was the kindest, cruellest, strongest, gentlest, cleverest, dumbest, most considerate, and most mischievous creature I had known. The past year had been more demanding than I had ever expected, but it was also the most fulfilling period of my life. I was totally committed to Cody's education and general well-being, and I could not see it ending, and I did not see why it ever should. After all, chimpanzees had been home-raised and kept with their adopted parents for more than six years. Why not an orangutan?

On the day of our departure the leave-taking went very well indeed. Too well really—there did not seem to be any inkling in Cody's mind that this was anything other than a normal day. His "kuhs," "puhs," and "fuhs" were as faultless as ever, and he even managed to piddle on my trousers, a sure indication that all was well with the world. I delayed as long as I could, tickling and playing with my little brown baby, kissing and being kissed. Then I put him in his bed and closed the iron door that separated the play and work rooms. Cody left the bed immediately and climbed the door, whining once or twice about my departure, but not more than usual.

What would become of him, would he pine while I was away? Would he still remember his words—and how to use them correctly when I returned? But perhaps Liz was right. As she had said, when we returned from Barbados and took Cody to Durham, I would have all the time in the world to work with him. He did seem to be learning new sounds very quickly now, and even if he did regress during my absence, I could surely bring him on much further during his stay in Durham.

But, I thought despairingly as I opened the outside door, no one really knows how a baby orang will react. "Perhaps you'll forget me completely," I said aloud. But Cody was not listening, he had become totally involved in attempting to remove a bolt from his side of the door, his long lips pursed in concentration as he pulled and plucked at the intriguing piece of metal. It was obvious that Cody had no idea I would not be with him for more than a month. I have always felt guilty about that; it seemed as if I was sneaking away under false pretenses.

18

He had hardly changed at all. Certainly Cody seemed no bigger than when I had left him. But watching him playing on his old broken-down chair and box, jumping blithely from one to the other and from there to the door, it was obvious that both his muscle power and coordination had improved immensely over the four weeks I had been absent. We had only just arrived at the zoo and, completely unaware of our presence, Cody was playing away happily, utterly absorbed in the intricacies of his self-invented game. The idea seemed to be to get from door to chair to box and back to the door in the fastest possible time. Cody swung from one object to the other with the easy grace of a professional acrobat, stopping occasionally to pound on the door or wall with his bent wrist when he felt he had done particularly well. As far as movement was concerned, Cody was years ahead of any human infant.

I suddenly felt very proud of my little toddler—for he was certainly no infant now—clambering, leaping, pounding, as fit and healthy as any youngster I'd ever seen. However, equal to my pride was a gnawing apprehension as to how Cody would respond at our reunion. Would the easy rapport of our first year be gone, replaced by the mischievous, yet basically uncaring attitude he displayed to strangers? And how complete would his "talking" be?

I moved slowly closer to the large glass window of his cage, raising a heavy bunch of driving keys level with my head. Still Cody played on, ignoring me—if he saw me at all—as just one of the many spectators who regularly attended on his antics. I shook the keys softly.

Cody swung his gaze indolently toward the source of this mildly interesting jingle. His gaze met mine and at once the eyes widened, bulging like large opalescent pearls while he stood stock still on the remains of his chair. He was as expressionless and immobile as a small red rock. The suspense was agonizing, and in order to break it I hastened to open the door and get closer to him.

On entering I found that Cody had moved the chair into his bed. Here he sat, playing quietly with his teddy bear and giving not the slightest indication that he was aware of my presence. It was as if I had never existed. I called to him through the dividing door. No response: the little ape simply ignored me, apparently engrossed in his play. It was horrible, had he forgotten me already? I went into the playroom and stood not four feet from him. Cody kept pulling and tugging at Freddy for about another 30 seconds.

Suddenly he stopped, let go of his toy, and stared straight at me. His face held an expression of such complete and utter melancholic bewilderment, so full of silent pathos, that I was lost for words. For what seemed like hours, we simply stared at one another. Then, to my relief, came bursts of high-pitched "pip-pirrups," Cody swarmed out of his bed and scooted across the floor on all fours toward me. Even before he reached me there came the familiar "pick me up" sound. I did not need to be asked twice. In an instant he was at my chest, his hairy arms clasped affectionately around my neck, blinking lovingly into my face and . . . Oh Lord! He'd gone and greeting-piddled me! I ignored the interruption and the warm, wet, glow down my front. I couldn't worry about something as minor as wet pants in the face of such a sincere demonstration of pleasure at my return. Cody had remembered "mama," even to the extent of bestowing upon me the traditional welcome. But best of all he had remembered the "puh" sound!

I sat down soggily on the floor and, in the hope of relaxing its desperate embrace, offered a brush to the hairy millstone that was attached with such frantic fingers to my neck. It took some minutes, for Cody seemed incapable of ever letting go, but gradually, with many a soothing word and gentle caress, he calmed down. I am quite certain that Cody's total disregard of me was because he had decided after the first week of my absence that I would never come back. The young ape had reconciled himself to the loss of his "parent" (much as captured orang infants do); and when I did reappear it seems that Cody did not want to admit my existence immediately lest I vanished once more. Well, I thought, he need not worry about my leaving again. I would give him all my attention from now on.

Ten minutes later Cody was sprawled out on my lap, as limp as a slice of liver, drinking in the sheer ecstasy of a long forgotten hair brush. His powers of adaptation were quite admirable: he even managed a quiet "puh" when I tried to stop combing him. Already it seemed so natural to be together. I could hardly believe I had been away. At feeding time it was the same story: Cody's "kuh" and "fuh" were very much in evidence. It was a huge relief to know that he had not regressed linguistically, and I offered up a silent gift of thanks to Nina and the other keepers who had kept Cody's words so firmly in his mind. After the meal, and throughout the rest of that day, I simply could not get rid of Cody. He clung to my chest the whole time as if in mortal dread that mama would abandon him again. Every attempt to place him on the floor was met with a flurry of desperate "puhs."

Two days later I did leave Flamingo Park Zoo, but this time Cody went with me. The zoo directors had honored their promise to let the infant move in with Liz and me at our home in Sunderland. We spent the last evening at the zoo feverishly packing our new acquired car, an old Morris 1000 about whose roadworthiness I had serious reservations.

Cody was loath to let us leave his room while we were packing, blasting out his brain-boggling screams each time we attempted an exit. As he would be facing quite an unsettling time when we left for Sunderland that night (we planned traveling overnight to avoid detection by prying neighbors when we arrived), I decided to indulge him. The work was doubly hard as I carried him wrapped in a blanket against the cold, from car to cage and back again, while we stowed whatever movables might remind him of his room at the zoo into our mobile cage. Freddy the teddy, rattles, A-frames, a favorite old shoe and several senile potties went into the trunk, and we ended by tying his beloved bed on to the rust-covered roof-rack. Into the car itself went all his food, a baby bottle, spare diapers and two flasks, the larger filled with orang-quieting milk and sugar. At last, with the deep bass of the village clock sounding midnight, all was ready for our departure.

Cody made everything much easier by obligingly falling asleep almost as soon as we settled ourselves in the car. Then we were off. The dank Yorkshire fog thickened appreciably as we proceeded peacefully northward. Crossing the Tees into the County of Durham at 4:35 A.M., we were reduced to crawling pace by its dense, cloying whiteness, which obscured all but five feet of road.

For the next two hours we moved steadily, if slowly, toward our destination. Then, only an hour from home came trouble. Cody awoke.

Now the little fellow was a handful at the best of times, but his most active period was ever the early morning, after a few restful hours in the Land of Nod. Our peace was shattered with the first flickering of his eyelids, and the back of the car sounded repeatedly with innumerable oohs, aahs, and screams from Liz, as she tried to restrain the mighty midget from expending his surplus energy on the leather seats and her blonde, plaited hair. Through the rear-view mirror I could see the two long red-haired arms falling time and time again on the face and tresses of my newly-wedded wife. How would she cope? It was yet another problem to worry over as I tried to follow the curb through the cotton clouds of fog. Was wife-beating by an orang grounds for a divorce?

After about an hour, Liz was more than a little disenchanted by our wayward passenger. I tried to tell her to hold out a while longer, as we were by this time less than two miles from home, but it was useless. Liz can take a lot—more than I can on occasion—but there comes a point when she decides she has had enough, and a steel screen slips down across her mind. Then, as I have found to my cost, her tiny foot goes down, stays down, and she refuses to pander to anyone. It was Cody's turn to learn. As he struck again her Barbadian accents swelled ominously.

"If yuh move again yuh goin' t'mek a wash uh trouble fuh yurself, hear?"

Whap! Whap! I could almost see Cody grinning as he made two more demoniacal leaps toward her tempting, pigtailed hair.

Whap! And a long, whining scream from Cody. I slowed down so far the car stalled, I cast an anxious look behind, expecting to see a pool of blood and the cadaver of one very dead orang. Instead, Cody was clinging fiercely to Liz, seemingly for reassurance after so abrupt and unexpected a chastisement. Liz sat expressionless, still holding a heavy hairbrush in her hand.

"He'll be OK now," she said impassively.

And she was right. Cody caused no trouble after that but what did act up was the car. It came to a sudden halt and adamantly refused to re-start. Each time I turned the ignition the headlights became progressively weaker. We were in a very worrying position. Every few seconds a car would zoom out of the mist and flash by at an alarming speed. It was infuriating to think that, only two miles away, my parents were sleeping blithely in their beds, completely unaware of our predicament. I dared not leave Liz and Cody and go for help because there was a real danger that a passing car might smash into our car in the thick fog. Should I try to stay and push it then? Perhaps, but I very much doubted my ability to move it. And if Liz and Cody were to get out into the cold night air there was a good chance that the little chap would catch a chill that could be fatal.

As I tried to decide what to do a large Ford skidded violently by. The driver, a huge bull of a man, emerged from the car and, swearing volubly, advanced angrily toward our broken-down heap.

"Bloody fool," he bellowed, "what the hell are you doing parked here like . . . What the hell is that?" he yelped, leaping back a pace, his hands held before his face for protection as Cody squashed his face suddenly against the car window. His anger quenched by an overpowering curiosity, the man listened like some big-eyed Friesian while I hurriedly explained our predicament.

Our new-found friend was so taken with Cody that he offered to tow us home, a proposition we regretfully accepted. Once again I marveled at Cody's magic influence, a unique ingredient in his personality that could turn both anger and fear to benign goodwill. We were lucky he was with us.

However, the tow from our Good Samaritan was not without its disadvantages. He dropped us at the edge of the green that fronted our home. We had planned to arrive incognito during the small hours and creep unseen into the house. Now here we were at 7:30 A.M.—just the time most people were waking or leaving for work—with more than 30 yards to cover across the green to the sanctuary of the house. Somehow, we had to transport Cody undetected across the open ground, but how?

The first step was to get Cody into his carrying position around my neck. I held out my hands to him, but for some reason he refused to leave Liz. It was hopeless: he clung grimly to his chosen perch, his four arms clasped tightly around her trunk, while I pulled and tugged in a vain attempt to dislodge him. Remembering Liz's angry hair-brush attack, I began to think Cody must prefer the violent type. So Liz would have to carry our hirsute hindrance after all.

Glancing around surreptitiously, I got out of the car and opened the rear door. As Liz emerged, her pongid impediment still clinging tenaciously about her belly, I quickly swept her voluminous coat around her shoulders, and with trembling fingers fastened up the front buttons. Although Cody was completely covered, Liz now possessed an ominous and altogether pregnant swelling beneath her wrap.

We started across the green, but had not covered more than six yards when Mrs. Jennings, from the next house but one, opened her door and stepped out, accompanied by a friend. We hurried on, trying desperately to retain an air of happy unconcern. But there was no avoiding it, we were going to bump straight into them.

Though I wished them a cheery good morning, I could see both pairs of eyes riveted on Liz's bulge. It was an enormous relief to step inside the house and hide ourselves from those prying faces. Closing the door, I heard Mrs. Jennings' high-pitched, whispy voice.

"There you are, Mary," she breathed, her voice edged with triumph, "didn't I tell you they got married a bit *too* fast?"

Having been awake now for more than 32 hours, we managed to palm Cody off on to my mother and retire to grab a few hours' sleep. My last recollection was of him leaping gleefully upon my bag of notes and drawings. At that moment, he could have had my diaries had he wanted them. Sleep was all. But after my morning nap the apprehensions returned. Had he destroyed my papers? How had Mum coped? Everything was so quiet. Too quiet! I leapt from bed and scrambled anxiously downstairs.

The sitting room was a scene of pure domestic tranquillity. Ensconced happily on the fireside rug was Cody, his brown hair highlighted with gold from the glow of the coals. He was as alert as ever. A tired, though self-satisfied, Mum sat beside him, her arm around his shoulders. They were both watching television. So unprepared was I for this improbable scene that I sank onto the sofa. Mum looked across and winked happily.

"Isn't he gorgeous? He's been like this for the past half hour." She pointed down at the telemerized orang beside her. "He's fascinated by the TV."

When I had come into the room Cody had cast a sidelong

glance in my direction, seen it was "only him," and quickly returned to his study of the screen. I could see his eyes following the army colonel who, to the accompaniment of stirring martial music, was striding across the parade ground to review his troops.

In the weeks that followed we were to find TV a relatively fool-proof sedative whenever Cody's cavortings became too boisterous to bear. He especially loved cartoons and quickly learned to push the right buttons to change from one channel to the next, usually in search of Woody Woodpecker or Bugs Bunny. It was funny at first, but after a few days the channel-changing trick became a bit of a bore: Mr. Spock would be about to apply the Death Grip to yet another wicked little alien when a flash of brown fur would rocket toward the TV set and we would suddenly find ourselves watching Huckleberry Hound!

On this first day, however, with Cody happily preoccupied, our immediate problem was to prepare his room. Knowing the infant's penchant for dismembering objects, our bedroom would have to be cleared of every item about which there was even the slightest suspicion of detachability. That meant everything. To begin with, however, my books and papers would have to go. Those who know me would realize the virtual impossibility of such a task. I keep so many books, files, letters, and slips of paper that might conceivably "come in useful" that my bedroom floor often resembles a sea of stationery, with only the bed managing to stay afloat. And even this is occasionally swamped by an excessive wave of correspondence. Still, the giant wastepaper basket had to be cleared, so we wearily set to.

By early evening, the papers and all other impedimenta had been removed, and all was ready for the youngster's reception. Cody's bed stood in the corner of the room which most closely approximated his cage at the zoo—facing the window and opposite the door. His A-frame stood in the center of the room and, as the final touch, numerous toys had been scattered across the floor. It seemed to work, for Cody appeared only slightly worried by his sudden translocation. He played happily around his bed, squeezing through the vertical rails at the side, only to crawl back out again to repeat the procedure. I was pleased to see him so at home in his new surroundings, but it was early days yet. The real test would come when we left him alone in the room at bedtime.

In the afternoon I had an hour-and-a-half's play session with Cody. As well as tickling, talking, and play-fighting with the toddler, I also included games of a more educational nature. His responses to these would be comparable with those given by human infants of the same age and should therefore be of considerable psychological interest. Two were of special note: first scribbling, and second nesting and stacking a series of pans. As on previous occasions Cody failed abysmally on both tests. His interest in drawing lay solely in the epicurean delights of pencil-chewing. Not once did he show the least inclination to "take a line for a walk."

It was the same story with the pans. There were six of them, in diminishing sizes, made so that they could be stacked one on top of the other or nested together in a neat little pack. Each day of testing I diligently stacked and nested the pans for Cody's benefit, stopping often to await his response. Cody always ignored the nest but he responded to the stack—with wild and malicious monotony he sent it flying! Cody was a year and three months old now, and past the age when young humans begin to show an interest in such things. Cody's lack of interest in the pans surprised me, for he had passed a similar "cube in cup" test in which the infant, presented with a cube and cup, has to grasp the cube, hold it over the cup, and drop it so that it falls into the cup. The test is designed to monitor, among other things, the child's growing awareness of spatial relationships. Cody had accomplished this maneuver (very difficult for any young primate, involving as it does a multitude of hand, arm, and eye coordinations) at 11 months of age—well before the time set down as "normal" for the human child. So I had been eager to see how the young ape would fare with the pans: but up till now his progress was nonexistent.

I looked at my watch—7:15 P.M. Another hour and I would be

putting him to bed. However, at that moment thoughts of repose seemed to be far from Cody's mind. He was still playing in his bed, this time lying flat on his back, the security blanket draped across his skyward-pointing limbs, and biting gamesomely at its well-frayed edges. He looked so happy—please God he wouldn't cry when I left him.

It was a vain hope. I shifted my balance, moving one foot slightly toward the door, and Cody sat bolt upright. I hesitated momentarily from the inevitable pick-up, and to my great surprise and joy there came several perfect, and very desperate "puhs." It was the first time Cody had used this sound so far from its original setting—an indication that his words were not bound to any particular location.

Cody first snuggled in close, then raised his wrinkled face to stare directly into my eyes, at the same time molding his mobile lips into a funnel-like pout. I had seen infant orangs do this to their natural mothers as a form of greeting, and my heart swelled with tenderness toward my diminutive ape child. He was so trusting that I felt a pang of guilt at the thought of deserting him. But what could I do? He had to learn to sleep by himself.

After a further ten minutes of playing and cuddling Cody slipped from me, onto the blanket in his bed. He sat content, smiling upward until I slid the half-inch chipboard across the top of the crib to prevent his climbing out. Then the smile vanished and Cody went berserk. I somehow managed to hold the heaving board flat against the top of the bed and to ram home the four bolts I had fitted to hold it secure. As the fourth and final bar slid home, I noted happily that all Cody's noise and activity had ceased. "There," I thought to myself, "that wasn't so bad, he's settled down pretty quickly." Turning to go, my foot struck something soft. I looked down, and saw on the floor an orangutan just like the one in the bed. It *was* the one in the bed! As soon as the cover was fastened. Cody had found a ready escape route through the side railings.

With Liz's help I dragged in an old, empty and very heavy mahogany chest of drawers from another room, pushing it up against the exposed bed-side. In went our protesting pongid, the top was once again secured, and we stepped back congratulating ourselves on a job well done. Just then the mahogany chest groaned away from the crib and a hairy arm appeared through the rails, following rapidly by the remainder of Cody's undulating carcass. His strength was phenomenal!

Seated on my shoulder, Cody busied himself with an in-depth examination of my left ear, while I racked my brains for some other cunning ploy to incarcerate him. If only we could keep him in there for an hour or so, until he went to sleep. Perhaps a heavy meal? No. While I had no doubt it would be ravenously consumed, food seemed to have no effect either way on his activity. It simply increased the turnover of dirty diapers. But hold hard—a meal might be just the thing!

Two apples, a large bowlful of porridge, sugar, and a mashed banana, followed by a huge mugful of milk. Cody must have thought it his birthday as I presented them one after another to his ever-eager maw. There was hardly time to burp, and, best of all, he didn't even have to "speak" for even a single spoonful. And the more he ate, the more certain I became that Cody would indeed remain in his bed the next time we tried to say good night.

When he had finally consumed the last of his supper, I gave him ten minutes to let the food settle (it would only make matters worse if he were sick), and then once again we went through the bedtime ritual. Immediately the crib-top was in place, Cody rushed for the railings. His left hand went through, then his head, but that was as far as he got.

The big supper had done its work. Cody's belly was so distended that it was physically impossible for a mound that size to pass between the close-spaced rails of the bed.

"Betrayed by your own greed," I admonished him; but I don't think he heard. Cody was so astounded he even forgot to cry, and as we left he was still looking down in bewilderment at his bulging, traitorous tum.

19

Once Cody had settled into his new home, I began the task of teaching him his fourth word. This was to be a "thuh" sound, for which he would have to hold his tongue between his teeth and, once again, expel air through his mouth.

But before I could begin I had to decide what Cody's new word was to stand for. This was hard: As he grew older, Cody had developed an almost complete nonattachment to most objects. While this is highly commendable if you are a Buddhist, it is not really helpful in the learning of new sound patterns. All he seemed to care about was food, drink, and personal contact with his "parent." As he already possessed words for these three desires ("fuh," "kuh," and "puh"), it became increasingly difficult to think of suitable rewards that would produce the high motivation necessary for speech training. Inspiration struck, or rather bit, when I received a nasty nip on the leg from Cody because I had stopped brushing his belly. Cody's fierce response made me realize how much he enjoyed this particular hedonistic delight. Here, I thought, was the answer; brushing would be the next reward.

It was a very anxious and exciting time for me. If, as I suspected, the little ape had learned what was expected of him in the talking session, then he should continue the trend of faster learning with each new sound taught. But would this trend continue, and how pronounced might it be? Cody's response surpassed anything I had dreamed of. I began tuition for his new sound, and Cody mastered it in only three days! Even a month before I would have considered such a feat impossible. Cody's rate of learning stands in sharp contrast with the Hayes's chimpanzee Viki, who took months, and sometimes years, to learn her words. It seemed that I had hit on a very efficient way of producing voluntary vocalizations in the orang, and perhaps all the great apes.

There was even more exciting news on Cody's usage of the four sounds. I had recently analyzed the sounds Cody emitted for his various "rewards"; for example, I compared the number of times the infant said his "pick me up" sound as opposed to "kuh," "puh," and "thuh," when it was obvious that he wished for contact comfort. Amazingly, his responses each day were almost always more than 70% correct, a success rate unheard of in ape vocal responses. Even Cody's off days were understandable—his mistakes concentrated around the times a new sound had been taught, or on occasions when I had redefined his words (for instance, when "puh" came to mean "pick me up," and not an indiscriminate request for food). These were all occasions when the status quo was upset, times when it might be expected that the infant would become a little confused.

It is important to realize that the vast majority of Cody's vocalizations were not produced in a standard environment, that is, not in the talking box. In all, Cody requested food, milk, and contact comfort in no less than 14 different rooms at seven locations. He was therefore not limited to producing his sounds approximately only in familiar surroundings. Neither did Cody work only for me; in other words, he was in no way person-bound. He would respond with the correct word for at least three different zookeepers, for my mother, my sister, and for Liz. Nor was object limitation seen in that Cody would say his food word equally well for fruit, vegetables, biscuits, and meat. "Kuh" was used for milk and other beverages in more than three types of cup. This evidence seems clearly to reveal that Cody possessed a simple concept for food and drink and that he was able to link these object-classes with his words. These abilities have not been shared by any other ape to which vocal language has been taught. Viki, for example, was extremely situation- and stimulus-bound. Her foster "mother," Catherine Hayes, makes some telling comments on Viki's capacity: "At the age of 30 months, she . . . used only three words, and these not always appropriately. In fact some days she either refused or was unable to say anything." Cody, by contrast, was only 15 months old, had four words, and was using them daily, and to a very great degree appropriately.

Just as important to Cody's appropriate use of his sounds were the novel purposes to which he bent them. Such behavior gave me valuable insights into the working of Cody's still developing use of "language." One fact that I discovered quite early on in the research gave me more weight to my belief that Cody was using his words as *verbal signs,* in much the same way as Washoe used manual signing to communicate with her human compatriots. At the beginning of talking tuition, Cody was very fond of vocalizing spontaneously, as if playing with his words. But as time went on such vocalization decreased in number. When Cody was taught his third word, they stopped abruptly and were never heard again. Now spontaneous vocalizations are just the group of sounds that one would expect to decrease if the infant were beginning to use his sounds as verbal signs, that is, to indicate specific object/activity classes. Cody I believe, had realized that words had uses, and were not, as it were, to be squandered on the empty air.

As the infant's vocabulary increased, the uses to which he put his words, in self-taught novel vocalizations, continued to expand. Cody rarely bit anyone, but when he did he was usually slapped for it. Until he learned his fourth sound such chastisement had always been accepted with typical simian stoicism. But now, when I slapped him for a nip of the calf and then turned away to show how annoyed I was at his behavior, Cody ran around to face me, climbed on to my knee, and looking me straight in the eye made two "thuh" sounds, as if this would please me and assuage my anger. From the context of the vocalization and from Cody's obviously harassed frame of mind, it was impossible to believe that he was using "thuh" in its correct sense, that is, for brushing. I am convinced that he had, by himself, invented another use for it—that he said "thuh" to pacify me, in other words to say "sorry."

Again, one evening at dinner, Cody approached Liz after seeing on her plate a succulent piece of meat. Unlike wild orangs, who are complete vegetarians, home-raised orangs often find animal protein very palatable. Standing next to Liz, Cody said "food" in a loud and confident voice. Liz was feeling hungry too, so she ignored the request, even after it was repeated in the same assured tone. Cody, frustrated by this unexpected turn of events, then angrily gave vent to his whole series of words, "kuhs," "puh," and "thuhs." This was a reaction that I saw only two or three times during the study, and it seemed to appear only when Cody was frustrated in an attempt to obtain something he particularly desired. It seemed that on these occasions, having failed to gratify his desires by recourse to the correct word, Cody resorted to a "quantity not quality" stratagem. All his sounds were tried out in the hope that one might weave the spell that would give him what he wanted. When Liz ignored even this exposition of his vocal art, Cody gave up public speaking and bit her on the wrist! Liz could not really ignore such determination on Cody's part: she asked him for his food sound again, Cody obliged and made off happily with a fine piece of steak.

Cody even seemed to invent new sounds. My mother reported that one day while she was orang-sitting, Cody boldly crawled over and began pulling her from her easy chair. He then proceeded to drag her into the kitchen so that he could ask for food. What was very unusual about this episode was that Cody produced several "sheesh" sounds while pulling his unwilling victim along. It was as if he wanted a word for "come follow" but, lacking such a verbal sign, was forced to make up his own.

This "sheesh" must have been the same one with which Cody

had serenaded me during that rather eerie episode when he had sat on my knee studying my mouth as I talked and made such an effort to talk back. It seemed to be a sound that Cody preferred and, as there was no question of his ability to produce it, I decided to make this his fifth sound as soon as the fourth had been consolidated. It would signify—as Cody had shown he wanted it to signify—"come follow."

As time progressed our little talking ape began to use his simple functional language to insist on what he wanted, even when it was out of sight. This began during Cody's fourteenth month of life during a feed session. Cody refused his last spoonful of cereal and instead uttered his "milk" word. I hesitated, and Cody repeated "milk" in louder tones. I was intrigued as to what the young creature would do and placed him on the floor at my feet, whereupon Cody made straight across the room to his cup of milk. Was he trying to tell me he wanted no more solid food while at same time indicating what he preferred? Was he asking for his milk? I was not certain that he was, in fact, capable of such communicatory skill but a second incident convinced me.

Cody was playing on the floor when I approached with a pan of porridge and honey, a special treat that the young orang really enjoyed. The following "conversation" ensued:

Cody: "Pick me up."
Me: (having expected Cody to utter his usual "food" sound) "No. What do you say?"
Cody: (halfway up my trunk) "Pick me up."

Cody then reached the upper portions of my chest, and settled himself comfortably on my right hip. Only then did he turn and, looking straight at the plate of porridge, say "food."

One strange aspect of Cody's vocal behavior was that, at first, he made no distinction between animate and inanimate objects. I have never read of a human child laboring under such a misconception, but Cody seemed convinced that he could get whatever he wanted by saying his words to a door, window, or even to a piece of wood. He first brought this to my attention early in his twelfth month. I was writing up my observations while he played happily in his bed with the portable test table, a laminate of plywood measuring three feet by two. The test table also doubled for a diaper-changing table when laid across the topmost rails of the bed, as it was on this occasion. Cody was pulling and pushing at the table, hanging from it with all his weight, or lying on his back and kicking at it so that the table jumped and leaped about the bed. Somehow he contrived to lodge one end of the test table in the crack between the end of the bed and the frame that held the side rails so that it stood up vertically. As far as Cody was concerned, it was now infinitely less playable-with. As I watched, he tried four times to pull the table back down to its original position. Each time he failed, and with each failure the little orang became more and more frustrated. Finally he released his grip on the table and sat back on his mattress. He could take no more. "Kuh," he commanded; and then, seeing no response, he rose and shook the board, at the same time voicing a loud "puh," By some strange chance Cody's shakings released the table from its position in the cot, and, obviously well satisfied with the efficacy of his magic spell, Cody pulled it toward him for a good chew.

It was probably Cody's success in releasing the table that led him to attempt even greater feats with his incantations. Having once sampled the delights of television, Cody was always eager to leave his playroom and get downstairs to see what was on the box. Unfortunately, a large door intervened. Cody was usually patient enough to wait until I opened this for him, but on one occasion he was so keen that he tried to open the door himself. Being still quite small he was hardly able to reach the handle, let alone turn it, and after trying once or twice without effect, Cody tried another tack.

He sat well away from the door, his back as straight as a die, and stared steadfastly at the handle. Then he voiced an imperious "puh" directly at the door, obviously expecting it to fly open. There followed a puzzled silence, during which Cody's eyebrows jigged about on his forehead like two demented caterpillars. Another commanding "puh" and another long silence. I could almost hear the wheels and cogs grinding their slow tortuous course as he tried to work out just what was amiss. His face brightened momentarily as the answer broke upon his tiny brain. Then his visage darkened and like some sorcerer conjuring unknown powers to his own fell purpose, Cody rose unsteadily to his feet, raised his fisted hand and shaking it high with tense frustration, called upon the door to open with a loud "thuh."

If there had been any justice the door would have instantly burst its hinges. When it did not, Cody looked around him, astonishment etched all over his face. What had gone wrong? I took pity on the little one after that. Not wishing to destroy his confidence in the power of words I lifted him to the handle of the door.

Living almost every day in the company of this baby orangutan made me very ape-orientated. I became quite out of touch with things humans, especially human development, and it was a lesson to me when my sister moved close to our house and was able to visit regularly with her 11-month-old daughter, Louise. The first time Cody and Louise met was a memorable occasion, and not just for them. Louise was sitting on the sofa when I brought Cody in, and for the first two or three minutes the reaction of both infants was muted; their faces remained neutral, and they simply looked each other up and down in a formal and ever-so-polite sort of way. Then Cody, who had been sitting on the floor, bounded on to the sofa and put out his hand toward the tiny human. Louise was a robust toddler and quite up to par as far as human infants were concerned, but in comparison with Cody's motor behavior she fell far, far behind.

Cody moved to the top of the sofa and again reached out to Louise; she responded in kind and their hands met. For a moment I was struck with the symbolic nature of this union, the way in which the orang had greeted its distant relation with that most basic of human gestures, the outstretched hand. Then Cody began to pull: Louise was dragged bodily from her seat and up on a level with Cody's face. Friendly he may have been, but Louise was no match for Cody's superior strength. We would have to be very careful that the orang-baby did not inadvertently use his powerful muscles to break the leg, arm, or neck of the human child. Strangely enough, with such a slow and obviously inferior creature, Cody never tried out his dominance routine on Louise. "The Chase" seemed to be reserved for more worthy opponents.

But if Louise was behind the motor development and strength, she far outstretched the orang intellectually. While it was true that Cody possessed as many words as Louise at this time, his ability had been earned at the expense of long hours of intense training. Louise, by contrast, had mastered her words with the minimum of training—it seemed that all she had to do was listen. In many other behavioral domains the story was the same; with pan piling for example, Louise would watch carefully as I constructed the tower of pans, and then attempt to do the same. Cody still took pleasure only in destruction.

It was after one of these piling sessions that, looking at the two of them side by side on the sofa, I suddenly felt very sad about Cody. He seemed to me the most unfortunate of creatures. The young ape's brain was almost as large as it would ever be. The orang is born with 65 per cent of its gray matter at birth, far more than a human. It was inevitable that as Cody grew, he would gradually fall behind in the race with human infants. It seemed to me tragic that Cody should be cheated of all that extra brain. But then I realized I was looking at my pupil's progress from an entirely human point of view. In orang terms, Cody was all right; he was built for an entirely different world from the one he inhabited. He was designed for the high canopy of a tropical rain forest, and, in his own niche, Cody would be a winner.

It was then that a small cloud of doubt began to darken the glow of my pride in all I felt I had achieved. Perhaps the real tragedy was that Cody was being brought up in an environment for which he was not suited.

20

All pongids eventually grow so strong they either become unmanageable, or an unwitting danger to humans. Cody, I began to realize, would prove no exception. Already, now he was 16 months old, it took me all my strength to restrain him when he decided on some forbidden course of action. I could not see myself asking Cody to say a new word when he was the size of Adam his father, much less give him a punishment period! I knew that he would have to go back to the zoo at some time, although perhaps not for some years, so periodically we went down to his old zoo hut in the hope that he would still regard it as one of his homes. It was on the second of these visits that something occurred to change the whole course of the research, and my relationship with Cody.

We had just finished lunch, and I was standing in Cody's room with him on my hip. A quick tug on the ear announced that Cody had finished his milk and was looking around expectantly for further labors worthy of his efforts. I picked up his wide, nylon hairbrush, and catching his eyes, slid to a sitting position on the tiled floor. Cody knew what was coming next, and he couched himself comfortably on my lap, legs crossed like some sybaritic sheik, as he prepared to give himself up to the pleasures of a prolonged brush and tickle. However, he now had to work for such sensuous delights, and it was not until I heard him utter "thuh" that I began, commencing slowly with the light delicate strokes across his belly. Cody was transported with rapture at the first touch of the bristle, his eyes glazed, his limbs limp, slowly turning his pliant trunk this way and that to ensure complete coverage of all his most susceptible spots. I allowed him 30 strokes, then stopped.

Impatient for his return to sensuous enchantment, Cody raised his head eagerly, mouth open.

"Thuh."

"Clever boy." Down came the brush again, and Cody instantly dissolved into luxuriating flaccidity.

And so we continued, 30 strokes, "thuh," strokes, on and on until Cody, tiring perhaps of such frivolity, began a new game, striking the soles of both feet against my chest, slowly at first, then harder, gradually working up to what promised to be an uncontrollable frenzy. More efficiently than any words, Cody was telling me he was bored with education. It was time to change tempo and begin something completely different.

Rising to a standing position, Cody clinging tightly to my chest, I looked across the yard to see Nina feeding the adult apes. It was one of those bright, lukewarm winter days when the adults were allowed out to exercise on the swings of their enclosures. Rather than call them back in for lunch, Nina was feeding them all through the bars, much to the delight of Janet, a newly arrived juvenile chimpanzee whom Nina carried everywhere with her. The adult chimps had already been fed, and Nina was giving Cody's parents, Adam and Mandy, their daily seven pints of milk each. This was the easiest part of the whole procedure for, unlike the fruit which had to be laboriously delivered piece by piece through the strong mesh bars, Nina had merely to point to the milk jug and an eager pair of lips poured themselves like liquid rubber through the mesh, re-forming instantly as fleshy funnels, milk-receptors. It was then a simple matter for Nina to pour direct from jug to jaw.

Her revictualing at an end, Nina climbed the spectator's barrier and moved around toward Cody's cage. I opened the door for our usual pre-lunch chat. We talked across the fence and with Cody on my hip and Janet on Nina's I felt like a gossiping Hausfrau hearing the latest scandal. Strangely enough, after a first initial touch, Cody and Janet seemed to decide to ignore one another, each spending the time grooming their respective "mother's" hair.

"Nelson was a bit down in the mouth today," Nina confided, "so I've given him a welly to brighten his day." Nelson had a curious fetish about wellingtons, indeed, anything made of rubber was sure to send him into paroxysms of sexual delight. Nina then told me that Cody's mother, Mandy, definitely seemed to be much more gentle today, almost maternal. This was good news. John Laing had told me that Mandy's always-distended stomach appeared even more protuberant of late, and that he thought she might be pregnant. Another young orang would certainly be a welcome addition to the zoo, and I might obtain valuable data on orang parturition. After Nina had left, I decided to take a look at the reformed Mandy. Seeing that it was such a fine day, I thought I would give Cody some fresh air too, and took him with me to see his real mother.

Carrying the little ape piggyback across one shoulder, I made for the orang section. Both parents were out in their exercise cage although I could recognize Adam only by his two legs which poked out incongruously from below his homemade "tent," a huge tractor tire angled into the junction of the back and side walls, and supported on either side by an enormous log. This primitive edifice was his newest weapon in the never-ending battle to avoid the unwelcome stares of visitors during his exercise period. Mandy, as usual, sat flush against the bars, her arms folded, and following my approach with sharp movements of her tiny brown eyes.

Cody was busy biting my neck as we drew close to the large

mound of red matting that was his mother. It was not until I had squatted down within two feet of the wire that he became aware of her presence. Mandy did seem to be very placid, but, much more important, I saw she was ignoring me, staring over to my shoulder to where Cody had only just raised his head from his busy job of chewing my sinews. She stretched out her open hand through the bars, her eyes more alive than I had ever seen them. Vainly, she tried to touch Cody's head.

Mandy's delight was not reciprocated. There was no intimidation display from Cody for this stranger. He shrank back in fear as soon as he saw her face, his spine freezing rigid in horror at the sight of the grisly ogress before him and his orange hair standing up all over his arms, head, and back. To my amazement, Cody was absolutely terrified at the sight of his own mother! He dropped bodily from my shoulder and scurried behind my legs, gripping them so hard that later I discovered four long, red weals on each of my calves. Here Cody huddled for protection like some cowed and shivering wreck, his eyes fixed resolutely on the ground in front of him.

I was thunderstruck by this ghastly turn of events—never had I expected Cody's rejection of his mother to be so violent, so absolute. At the same time it was a deeply moving scene. Looking from Mandy, still trying forlornly to reach her young son, to Cody, head down and trembling behind my knees, I came very close to tears. In a sense, this was all my fault: I was to blame for this pathetic little tableau. Swiftly I bundled Cody up in my arms, and with Mandy's hand still outstretched in silent supplication for her infant, I left for the sanctuary of Cody's cage.

After this, I had some deep heart-searching to do. It was true that Mandy's behavior could be put down to motives other than maternal love: she behaved in an identical fashion whenever she wanted food, or indeed anything of interest. But even with that segment of the episode carefully rationalized, Cody's abject terror of his own kind simply could not be so easily dismissed. It seemed certain that, having been raised in the company of humans, Cody had come to believe himself human too. He had, after all, seen almost nothing but members of the species *Homo sapiens* since the day he was born. If Cody now saw himself as more human than ape, how would he ever mate and produce children for his species?

The question was of more than academic or personal interest. The orangutan is an endangered species; it is found today only in certain small areas of Borneo and Sumatra and is dependent for its survival on a specific type of primary (or old secondary) rain forest, a habitat that is under sustained pressure from commercial timber cutters. There are now more people in one square mile of central London than there are wild orangs in the whole of the world. The exact figure is unknown, but not more than 4,000 wild orangutans survive, scattered in small pockets on these two Southeast Asian islands. The zoo population of this species is less than 500. In recent years there has been some success in breeding orangutans. I wanted Cody to participate in the attempt to save his species, but if this plan was to come to fruition then Cody would have to be returned to an apelike condition *before* his identification with humans became irreversible.

Cody was causing me some problems. I had recently been assigned a new supervisor, one who had taken exception to the "very unconventional" aspects of my research. The idea of Cody living with me at home filled him with all sorts of fears. What, he demanded, would the zoo do if Cody suffered some accident, if he was run over by a car or died of pneumonia? My supervisor was of the opinion that the zoo directors might be a little more than sad at Cody's untimely demise. They might sue the anthropology department for "the market value of the orang," which my supervisor had discovered was close to $7,500. He had decided that it was best for the department as a whole if Cody was returned to the zoo as quickly as possible.

Someone else who would not be unhappy to see Cody go was Liz. Although she had done and said nothing to hinder my work, I had noticed that since we were married she had become rather resentful of Cody. I could understand how she felt; I too had not imagined that our first months of married life were to center around the care and feeding of an infant orang, a creature equivalent to human triplets as far as energy and mischief were concerned.

On a purely personal level, I did not want a mixed-up neurotic orang on my conscience. Apes possess more than just a physical similarity to humans. My work had shown how similar the development of orang and human was over the first years of life, and others had shown that, psychologically, apes are capable of almost as many "feelings and emotions" as a human being. If denied a piece of food, the infant orang is capable of a temper tantrum indistinguishable from a human toddler. Left to themselves all apes turn morose and ill-tempered; and if you show too much affection to one of a pair of orangs the one neglected will certainly become extremely jealous. Even more to the point, I had already been told of one female orang who, shut up for six months in a tiny cage, found it impossible to relate either to humans or to her own kind when she was finally released. She ignored everyone and simply rocked herself incessantly, in exactly the same way as certain severely disturbed humans. Cody might easily suffer a similar fate, for when he did become too strong to control safely, there would be no alternative but to keep him caged, set apart from his human companions who might suffer serious harm from even the lightest of Cody's "playful" slaps. If, denied the fellowship of man, Cody was unable to relate to others of his own species, then the miserable life of a lonely, pathetic misfit seemed to be his inexorable fate. That destiny was one I was determined he should avoid.

There was, however, one grain of comfort among all this despair. Although Cody was definitely terrified by adult orangutans, I knew he was quite amenable to young apes, such as Nina's Janet. With the help of young apes, then, we still had time to pongidize him before it was too late. But clearly Cody would have to be returned to the company of other apes immediately, or he would soon become irreversibly prejudiced against his own species.

But what about my work? There was no argument really: my work would have to stop if it meant turning Cody into a frightened mixed-up neurotic. I had already shown that apes *could* learn a simple language. And I had enough data for two Ph.D. theses.

I simply could not inflict such suffering on the little brown bag of mischief that I had watched grow from no more than a foot in length, who loved me as his only parent, and whom I loved equally in return. I patted the head of my ape child, still clinging tightly to me, and only now showing signs of recovery from what had been a traumatic experience. Much though it pained me to leave him, and though it would no doubt upset him too, the alternative path would be far harder for Cody. To persist with my study would mean that I had denied him his one chance of happiness, all because of my selfish curiosity about his ability to learn language. It was better to leave him—and to do it now. I had first to ensure that Cody would be given the best possible treatment at the zoo—that went without saying—but then I must make as swift a break as possible.

Cody raised his head, looking for a word of reassurance.

"It's all right, Cody," I told him, my voice tight in my throat.

He relaxed on to my chest, uttering a soft, restful "puh."

21

Cody is three years old now, and doing handsomely. He still lives in the glass-fronted cage of his babyhood, but he himself is changing rapidly as the years go by. There is even a hint of the huge cheek flaps which, when fully developed, will mark him as an adult male. Cody is growing up.

Our parting, when I took him back to his hut for the last time, was for me a sorrowful affair. Cody seemed to know that something was amiss. Even though I tried to act as if everything were normal, Cody's behavior showed that he could feel the tension in the air. On the last day he was far more dependent than he had been for months, crawling over to me constantly and saying "pick me up" so that he could be close. He cuddled and snuggled for all he was worth, protesting strongly each time I tried to put him down. But I hardened my heart. While Cody was absorbed in a play fight with Freddy the Teddy, I slipped quietly out of the playroom and, closing the door behind me, slid the bolt into place.

At that noise Cody was galvanized into frantic action. He rushed to the door and climbed it, pulling vainly at the bolt. When he realized it was hopeless he hung dejectedly from one hand, his mouth one huge pout, crying softly and woefully through the wire. I think he knew I was abandoning him, and the knowledge that he knew made me feel all the worse. But it *was* for the best, I kept telling myself.

Now I go to visit Cody five or six times a year, and on each occasion he is less and less inclined to pine when I leave, especially since I arranged for the zoo to acquire a second infant orangutan, Little Dick. I think Dick's arrival helped to soften the blow of my disappearance. The two soon became inseparable, even eating from the same plate (a sign of true friendship in this greedy ape) and walking together hand in hand.

Each time I visit I note some new quality in Cody's personality or activities that brings me up short. He is, these days, much more aware of himself as an individual interacting with other individuals, rather than the egocentric little creature he was during his first year. This expansion of awareness is reflected in his ability to solve quite complex problems, and I am often tempted to try out particular psychological tests on him. But I have so far managed to restrain myself; my visits are purely social and my research with Cody at an end. Or nearly so. I am still trying to work out *why* he did what he did during his first year of life.

What is obvious is that the development study was a success. For the first time there is a relatively comprehensive account of the physical growth and the psychological development in this least studied of the great apes. Such data was lacking before and I hope it will now be of use to other researchers. Most important of all, perhaps, we now know that much more can be achieved in training apes to talk than had been believed before the research with Cody began. This particular orang was able, using an operant reward regime, to learn four sounds and to link them with objects and activities as verbal signs. In doing so, he not only shook established scientific opinion, he also raised the dignity of his own underrated species.

The chimpanzee's vocal linguistic achievements pale into insignificance against Cody's "language." The most successful chimp was Viki, with four definite sounds learned over a period of six years. Cody achieved a similar total in less than eight months. The quality of Cody's sounds was much higher than that achieved by Viki some 25 years previously, though the disparity may be due to the difference in training methods.

How the chimpanzee would fare on a regime similar to Cody's is an open question. The chimpanzee is far more social than the orang; it is closer, evolutionarily speaking, to man, and the chimpanzee's larynx and facial musculature are far more like those of *Homo sapiens* than are the orangutan's. In addition, Adrian Kortlaandt has drawn attention to the fact that, unlike the orang, champanzee infants babble in the same way as human infants. All these factors seem to indicate that the chimpanzee may produce as good or better vocal showing as the orang, should similar training be given.

Whatever the vocal competence of the chimpanzee, Cody's achievements have given insights not only into the mind of the ape, but also into the evolution of our own language. Haldane has suggested that the naming of objects and persons was the main reason why the vocalizations of our ancestors developed into the human language we know today. One can certainly see the advantage that naming would have for a race of hunter gatherers. But just when did such naming occur? All behavior is ultimately related to the brain, so the nub of the problem seems to be "at what level of brain complexity can a primate begin consistently to use voluntary vocalization to signify an object or action?" No matter how skilled in gestures, voluntary vocalization is the *sine qua non* of man's main communicatory system, vocal language. Cody's achievements in this regard show that, given a neutral matrix of equal complexity to a 16-month-old orangutan (and sufficient training) it is possible for the association of a specific sound to specific object to be formed. This may be regarded as a form of protonaming. As both the hominid genus *Australopithecus* and the early *Homo* species found recently possess cranial capacities whose volumes are equal to or greater than that of an adult orang, we should perhaps credit both these types with at least the rudiments of language.

Cody may, of course, be a genius, an Einstein of his race. It would be an extremely rash man who generalized on the basis of these results alone, and claimed such potential for *every* member of the species *Pongo pygmaeus*. But equally, Cody may be something of a dimwit, a dunce among his peers, or he may be "Mr. Average." We can do nothing but speculate until several orangs have been raised in a human environment and studied from birth to maturity. Perhaps, if the threat of extinction is removed from this species and the orangutan is secure in the wild, it will be possible to repeat the Cody experiment and to take training through to adulthood. There is still much to be learned from work of this kind. I have the feeling that the orangutan, seemingly the dourest and dullest of the great apes, has many surprises yet. We have only touched the surface of this ape's language potential.

But to speak about Cody "scientifically" seems to me to deny the essence of my experience with this baby ape. What appears just as important—and I confess that it is a very personal and highly biased opinion—is the rapport that developed between us during the period of research (and which I hope continues, though on a lower level, to this day).

It is difficult to describe the intensity of that very abnormal relationship—we lived cheek by jowl for so long that I felt I was almost part orang myself. I learned the truth of the old adage "language is often redundant." I could intuitively sense Cody's wants, his moods, as soon as I walked through the door into his cage. And I am convinced that he could do the same.

I demanded much from my young pupil: growth measurements, photography, interminable testing, and the long, hard slog of talking tuition were all thrust upon him whether he liked it or not. But I like to think that Cody also gained from the experience. The continual stimulation did, I believe, greatly develop his dormant intelligence and interest. It was a good time for both of us, a joyful, angry, bewildering, frustrating, fascinating time.

What will happen to Cody? He now appears to identify himself as "orangutan"; and after puberty (around seven years of age in the orangutan), provided his health remains as robust as it is now, there is an excellent possibility that Cody may hear the patter of tiny prehensile feet. *If* he meets the right girl that is. I can see her now. She'll be around four foot three, red hair, dark mysterious eyes, and with a 98-120-120 figure that will knock him flat. Only the best for Cody.